Laboratory Manual
for
Denise Guinn's

Essentials of General, Organic, and Biochemistry

Second edition

Julie Klare
Fortis College

W. H. Freeman and Company
New York
A Macmillan Higher Education Company

ISBN-13: 978-1-4641-2507-2
ISBN-10: 1-4641-2507-4

© 2014 W. H. Freeman and Company

W. H. Freeman and Company
41 Madison Avenue, New York, NY 10010
Houndmills, Basingstoke, RG21 6XS, England

www.whfreeman.com

Table of Contents

Introduction

Although a great deal of this material can be learned by reading and study, there is no substitute for a well-chosen lab exercise to bring it all together, regardless of the current level of understanding of the student walking into the lab.

- A student who is keeping up well and not struggling with the class will have the satisfying experience of being able to apply their knowledge, make predictions, and see how it all works out.

- Should a student have a beginning grasp on the concepts, seeing the activity unfold often leads to an "aha!" that brings it all together and allows full sense to be made of the concepts being demonstrated.

- A student for whom the concepts come less easily will be able to review an exercise that may not have completely made sense at the time, and see how the concepts learned in class explain what happened in the lab.

In all three cases, the goal is the same: perfect the understanding of the student as much as possible in a science field in which one concept builds upon another, and a missed idea may very well come back unexpectedly to cause a problem.

Once past the intermediate portions of the course, the labs will all have the same basic structure. A pre-lab set of questions and/or calculations will introduce the material in a slightly different context and provide questions designed to prime the students' understanding of what is to follow. The pre-lab is meant to be completed in advance, ideally before walking into the lab, and should be turned in before the lab is begun.

The pre-lab is followed by a protocol in standard format: Materials, Procedure, and Post-Lab Questions. These questions should confirm student understanding and can be turned in during the next class period or used as a basis for classroom discussion.

Perhaps most importantly, the lab provides a less-formal atmosphere than the classroom can provide, in which questions should be encouraged. There is not always enough time during lecture to ensure that every single student understands the material. In the lab, the students who need more time can get their questions answered while a policy of benign neglect by the instructor can be followed for students who understand well what they are doing.

The time in the lab should be an eagerly anticipated opportunity to try out what has been learned, cement an understanding of sometimes difficult concepts, and even have a little fun. This is a wonderful subject, and even a surface study of it will provide a deeper understanding of the natural world.

Introduction

Laboratory Safety

Working in a chemistry lab is not unlike working in a kitchen. For a novice, the dangers are not always obvious, which is what necessitates an introduction to laboratory safety. Your instructor will warn you about the dangers of each individual activity, which range from uncontrolled combustion to cold burns, but you are responsible for paying attention to your surroundings and abiding by the rules provided to you by your instructor. Some general chemistry lab safety rules include the following:

1. **Do not eat, drink, chew gum, or apply cosmetics in the laboratory.**
   Although the chemicals you may be using in a particular activity may not be toxic, you do not know what has been used in the lab prior to your arrival. Do not put anything that has been on the benchtop in or near your mouth (pencils included!). Before leaving the lab, wash your hands.

2. **Dress appropriately for the laboratory.**
   Open-toed and/or open-topped shoes, bare midriffs, and unsecured long hair can all pose a danger in the lab. Do not wear your favorite clothes, as they may be exposed to acids or oxidants. Contact lenses are not suitable for the laboratory. Safety glasses and gloves should be worn whenever directed by your instructor.

3. **Know the location and proper use of all safety equipment.** Perhaps the most important piece of safety equipment in the lab is the sink, which is the first place you should go to wash any chemical off your hands. If they are available, you must also know the location of the eyewash, safety shower, and emergency exit.

4. **Observe good housekeeping practices.**
   Having extraneous items in your workspace or leaving bookbags or other articles in pathways is not only a hazard for you but for other students as well. Clean up spills immediately: in the lab, a clear liquid left on the benchtop is as likely to be a strong acid as it is to be water.

5. **Dispose of all laboratory wastes appropriately.**
   Almost everything used in these labs can be discarded in the sink with copious running water. However, if you have not been told specifically how to dispose of something used in the lab, ask before pouring it down the drain.

6. **Ask for assistance with broken glass.**
   Your instructor will direct you to a dustpan and broom. Do not pick up pieces of glass with your bare hands. Do not place broken glass in the regular trash: place it in a designated glass disposal container.

7. **Be prepared for your procedures.**

   Before beginning any protocol, be sure you have read it to the end and are confident of how to proceed. If you have any questions, be sure to ask your instructor before beginning.

8. **Assume that all reagents are toxic.**

   Always err on the side of caution, particularly with unknown chemicals. Do not attempt identification by tasting or smelling chemicals. Waft fumes toward you using your hand to get a reduced dose of the fumes.

9. **Inform your instructor of any allergies or medical conditions.**

   Allergies, particularly to latex and sulfur, can be problematic in the lab if your instructor is unaware of them. Also, inform your instructor of conditions such as epilepsy or pregnancy to ensure that proper precautions can be taken.

10. **Do not engage in horseplay.**

    Any activity that your instructor feels poses a danger to you or to those around you will result in expulsion from the laboratory.

11. **Never walk away from a heating source.**

    Whether it is a burner or a hotplate, it must be supervised at all times whenever lit or hot.

12. **Do not look into a container that is being heated.**

    Should the heated substance pop or spit, damage to the eyes or face is likely to result. Also, be careful where you are pointing a test tube that is being heated.

13. **Stick to the protocol.**

    The lab activities are safe when the instructions are followed. If you choose to disregard the instructions, you are putting yourself and your neighbors in danger.

14. **If you have any questions, ask!**

    Your instructor wants you to have the best possible experience in the laboratory and learn as much as possible. If you are uncertain about any part of what you are doing, get clarification as soon as you can.

Laboratory Safety Quiz

1. When you first walk into the lab, you notice a spill on your portion of the lab bench. What is the best way to proceed?
    a. Put your bookbag down in the spill; your bookbag is nylon, and water won't hurt it.
    b. Ask the instructor what the spill is likely to be, and proceed according to instructions.
    c. Get some paper towels, and wipe it up.
    d. Dip your finger in it, and taste it.

2. When your lab partner arrives, you notice that she is wearing a very cute pair of flip-flops with chemical symbols on them. You should:
    a. compliment them.
    b. ignore them.
    c. ask if she has a pair of sneakers with her.
    d. leave the lab. You shouldn't have to work with someone like that.

3. The most important and frequently-used piece of safety equipment in the laboratory is the:
    a. eyewash.
    b. safety shower.
    c. fire blanket.
    d. sink.

4. If your instructor sees that you are jokingly trying to pinch your lab partner with a pair of hot tongs while trying to step on his feet, she will likely:
    a. expel you from the lab for the remainder of the period.
    b. ignore you; youthful hijinks are a part of the school experience.
    c. ask you to use cold tongs instead.
    d. compliment your lab partner on his sturdy shoes, as they are protecting his feet.

5. Your lab protocol requires that you heat a solution on a flame for 20 minutes. You are really thirsty, so:
    a. you and your lab partner go down the hall to get a soda, and come back in 20 minutes.
    b. you get a drink while your lab partner runs to the bathroom. You've got 20 minutes!
    c. you ask your lab partner if you can alternate watching the flame so it is never left unattended, and you can both step out of the lab for a few minutes.
    d. you ask your instructor to watch your burner for you. It's her job!

6. Your protocol requires you to choose an unknown from a selection of four white, crystalline powders and determine its properties. You:
    a. treat it like it is terribly poisonous. After all, it's *unknown*.
    b. take a quick taste to find out what it is and speed up your work.
    c. assume that it's nothing dangerous: all the unknowns look like either sugar or salt.
    d. mix a little of it with your neighbors' unknown to see what happens.

7. Your uncle and your sister both have severe latex allergies. You should:
    a. not bother your instructor with family stories.
    b. let your instructor know so that she can provide guidance.
    c. refuse to participate in laboratory activities.
    d. don't worry about it, as it will probably be fine.

8. You drop a beaker on the bench at a funny angle, and it shatters. You should:
    a. immediately start to pick up the pieces.
    b. just leave it. The janitor will get it later.
    c. ask your instructor how to proceed, if the sound of breaking glass didn't already bring her ~~running~~ walking quickly to your bench.
    d. tell your lab partner to clean it up.

9. There was a part of your protocol that didn't really make sense to you. You should:
    a. not worry about it. It will probably make sense when you get to that part.
    b. ask your lab partner if he understood that part of the protocol.
    c. ask your instructor before you start to make sure you don't make a mistake.
    d. B and/or C, just make sure it makes sense before you start.

10. After you are done testing your unknown, you are pretty sure you know what it is and it's time to get cleaned up. You should:
    a. put the rest of your unknown in the trash before you leave.
    b. ask your instructor for guidance on waste disposal. She probably just forgot to mention it.
    c. pour the rest of your unknown in the sink and rinse it down the drain.
    d. leave everything out on the benchtop and go.

11. Lab is in August, in the afternoon, and you have to walk a quarter mile from the bus stop. You should:
    a. bring appropriate clothing with you and change when you arrive.
    b. not worry about it: your instructor will understand why you are wearing sandals and a cropped tank top in the lab.
    c. keep a pair of socks and sneakers in your bag and get yourself a lab coat.
    d. A or C.

12. It is okay to keep your bookbag:
    a. in a cabinet.
    b. on the benchtop.
    c. in the pathway between benches.
    d. under your lab stool.

13. You are using an acid, and some of it gets under the cuff of your glove and onto your skin. You should:
    a. get that glove off and head straight for the sink.
    b. find your instructor and tell her.
    c. go look at the MSDS.
    d. don't worry about it; acid burns only happen in the movies.

14. You are required to heat a test tube full of liquid. It is important not to:
    a. point the open end of the test tube toward another lab group.
    b. put your face over the test tube to see how it is going.
    c. A and B.
    d. wear your safety goggles.

15. Unsecured long hair is dangerous in the lab because it:
    a. is distractingly pretty.
    b. can brush through chemicals and set on fire.
    c. shields you from fumes.
    d. can be used by others to clean up spills.

Introduction

Specialized Laboratory Equipment

Working in the lab will require familiarity with some equipment that is not commonly used anywhere else, and the purpose of these various items in not always perfectly clear to a beginner. What follows is a brief introduction to the specialty items that are used in the labs that follow, along with a brief description of their use. Be sure to familiarize yourself with these because it is normal for there to be a quiz on laboratory equipment before you are allowed to use it.

Beaker: This container comes in many sizes and shapes and is usually heat- and flame-proof. Don't be deceived by the volume markings. This is NOT a measuring device, just a vessel for holding things. Those markings provide only a rough estimate of volumes, generally +/− 5–10%.

Erlenmeyer flask: This is also a container, but one that is designed to minimize splashes and spills. It also has estimated volume markings, but it should not be used to measure volumes. If it is necessary to determine the volume of a flask, other equipment will be needed.

Graduated cylinder: This is a true piece of measuring equipment. If you need to measure a volume, this is a good thing to use. It can be used to take displacement measurements of irregular objects small enough to fit into it, as well as providing the volume of liquids or allowing precise quantities of liquids to be measured out.

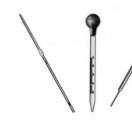

Volumetric pipets, graduated pipets, and droppers: A volumetric pipet is calibrated to deliver a specific volume—and only that volume. A graduated pipet can be used to measure a range of volumes, and a dropper can be used to deliver a small but indeterminate quantity of liquid.

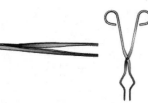

Forceps and tongs: The outsized tweezers used in the lab to carefully pick things up that are too small to be handled easily, are dangerous to touch, or need to be protected from contact are known as *forceps*, while the larger equipment used to carry hot glassware is *tongs*.

Introduction

Reading and Understanding Material Safety Data Sheets

The U.S. Government's Occupational Safety and Health Administration (OSHA) is responsible for the Hazard Communication Standard 29 CFR 1910.1200. The purpose of this standard is "to ensure that the hazards of all chemicals produced or imported are evaluated, and that information concerning their hazards is transmitted to employers and employees. This transmittal of information is to be accomplished by means of comprehensive hazard communication programs, which are to include container labeling and other forms of warning, **material safety data sheets** and employee training."

The Hazard Communication Standard (HCS) specifies 16 required elements that must be included on a material safety data sheet (MSDS). These are: identification, hazard(s) identification, composition/information on ingredients, first-aid measures, fire-fighting measures, accidental release measures, handling and storage, exposure controls/personal protection, physical and chemical properties, stability and reactivity, toxicological information, ecological information, disposal considerations, transport information, regulatory information, and other information.

Although the Standard requires this information, it does not provide a required format. The American National Standards Institute (ANSI), a private nonprofit organization, provides a standard that is encouraged by OSHA, but not required. As a result, MSDS can look very different.

In a health care setting, there will be numerous chemicals, from reagents to sanitizers to cleaning supplies. For all of them, the manufacturer provides, and your employer must make available, an MSDS. An MSDS contains a great deal of information, but it can be difficult to quickly find the information you are looking for without an understanding of the way these forms are set up.

When you have forms of the two types, it will be immediately obvious what their differences are. Some will contain all the information on one sheet, while others will run six pages or more. The length of the MSDS does not in any way correspond to the toxicity of the substance: you can find a short MSDS for a toxic cleaner, and a tremendously long one for table salt.

What you are usually looking for on an MSDS is health hazard data. This can take two forms: acute or chronic. Acute exposure is a sudden large exposure. Chronic exposure takes place over a length of time. For an acid, for example, an acute exposure might result in burned skin, while chronic exposure might result in lung damage from breathing the fumes over and over again. Treatment for acute exposures will also be given. Often, this is as simple as washing hands and moving to fresh air. Particular care must be taken in cases of ingestion: inducing vomiting often seems like the obvious course, but there are substances that can do as much damage coming back up as they did going down.

A measure of toxicity that you may see on an MSDS is the LD_{50}. This test was developed in 1927 and phased out in 2000. It is defined as the dose of a chemical that when given to a group of experimental animals will cause 50% of them to die. The animal type and dosage route will also be given.

MATERIAL SAFETY DATA SHEET (MSDS) ACTIVITY

Choose a substance of interest. This can be just about anything from an element to an individual compound, a medication, or a complex mixture like a cleaning solution. Search online for the MSDS for that substance. You may need to be flexible. If you do not find it quickly, go on to something else. Using a printed copy of your MSDS, answer the following questions (write N/A if the information cannot be found). **Highlight the source of the information on your printout.**

1. List two synonyms for your substance.

2. What is the LD_{50} for your substance? How was it determined?

3. There is some question as to whether an LD_{50} value can be extrapolated to humans, but if it could, what would be the dose of your substance that would kill 50% of 80 kg humans? (If no LD_{50} was listed for your substance, use 80 mg/kg. If more than one LD_{50} is listed, as they can be calculated in different ways, choose one.)

4. Does a higher LD_{50} indicate a compound is more or less toxic? Explain.

5. What is the most toxic route of exposure? What is recommended in this case?

6. What are the chronic effects of exposure?

Chapter 1

Measuring Matter and Energy

For scientific purposes, measurement and calculations are done in a very specific way that is somewhat different than the way measurements are done to build a house or bake a cake. The measurements and units may be the same, but the numbers are treated very carefully to minimize the *propagation of error*.

When a measurement is made, there will always be some error. The better the measuring equipment is, the smaller the error will be. Imagine a measuring stick with markings only in centimeters, and another with millimeter gradations. In measuring an object with the first measuring stick, the measurement will only be accurate to the nearest centimeter. The next digit will be an estimate and will contain some error. If the same object is measured using the second measuring stick, the measurement will be accurate to the nearest millimeter, with the next digit an estimate. The measurements of the same object might look like this:

10.6 cm with the first measuring stick. The 0.6 is an estimate.

10.54 cm with the second measuring stick. The 0.04 is an estimate.

If these were cubes, and you needed to calculate their volume, you would cube these numbers. A calculator will give you:

1191.016 cm^3 for the first

1170.905464 cm^3 for the second

It should right away be clear that these numbers cannot be used as they are, so to make them more reasonable, significant figures rules can be applied:

1190 cm^3 and

1171 cm^3

The difference in these numbers can be attributed to the propagation of error: there are small errors in both numbers, but when they are multiplied together, the errors are also multiplied.

To a certain extent, this is unavoidable. In scientific work, care is taken to simplify the way measurements are taken to reduce errors of this kind. A simpler measurement will always be better. To give a slightly absurd example, if you had to measure the distance between two mountain peaks, it would make more sense to use a precise detector and a laser to time how long it took the laser to travel from its source on one peak to the reflector on the other and back than it would to station someone on the top of each peak and time how long it took the sound of an air horn to travel from one peak to another by counting "one hippopotamus, two hippopotamus,...." Using the wrong equipment can lead to big problems: the scientists at CERN in Switzerland once excitedly reported that they had detected a faster-than-light particle only to realize that in a facility with billions of dollars worth of fabulous equipment, a slightly inaccurate measurement had been taken by equipment with a loose connection by subtracting out a measurement of the speed of light performed.

This story illustrates far more than just the fact that using the right equipment is important. It also shows how exciting science is to those who love it, understand it, and pursue scientific exploration right to its limits. There are not many people willing to spend large chunks of their lives underground looking for invisible particles, but those who do can be excused for their bursts of excitement and enthusiasm when they see something unexpected, even if it was because they were using an air horn instead of a laser as their measurement device!

1. Using dimensional analysis, make the following conversions (show your work):

    a. 18 inches into feet

    b. 18 inches into centimeters

    c. 18 inches into meters

    d. 18 inches into millimeters

2. Given the following measurements, calculate the density of the following materials.

    a. Dogwood block, 2 cm x 5 cm x 3 cm has a mass of 22.5 g.

    b. Ebony block, 10 cm x 5 cm x 8 cm has a mass of 480 g.

    c. Bamboo block, 3 cm x 4 cm x 7 cm has a mass of 33.6 g.

3. You are given 30.0 mL of an unknown liquid. What will the mass be if it is:

    a. ethanol (density = 0.789 g/cm^3)

    b. water (density = 1.000 g/cm^3)

    c. dichloromethane (density = 1.326 g/cm^3)

4. Given the densities calculated for the woods, and the densities given for the liquids, will the woods sink or float in each of the three liquids?

    a. Dogwood block, density =
        i. Ethanol –

        ii. Water –

        iii. Dichloromethane –

    b. Ebony block, density =
        i. Ethanol –

        ii. Water –

        iii. Dichloromethane –

    c. Bamboo block, density =
        i. Ethanol –

        ii. Water –

        iii. Dichloromethane –

MATERIALS

Meter sticks marked in both inches and centimeters

Ruler, metric

25 mL graduated cylinders

Lab manual

Glass or plastic and metal marbles

Calipers

Water

Density unknowns

Balance

Burner or hotplate

Specific heat unknowns

Beaker large enough to hold all the specific-heat unknowns

Beakers just large enough to hold one specific-heat unknown

Tongs

Thermometers

Volumetric pipets, 10 mL or larger

Pipet bulbs

50 mL beakers

Distilled water

Diet cola

Regular cola

Lemon-lime cola

90% isopropyl alcohol

200 mL graduated cylinders

Corn syrup

Food coloring

Vegetable Oil

Cork pieces

Ice cube

Pebbles

Chalk

Paper clips

Aspirin

250 mL Erlenmeyer flasks

Balloons

Dry ice

PROCEDURE 1.1: MEASUREMENT

1. Use the ruler to measure the length, width, and thickness of your lab manual in inches. Record your results.

   Length _____in   Width _____in   Thickness _____in

2. Convert these measurements from inches to centimeters (1 inch = 2.54 cm).

   Length _____cm   Width _____cm   Thickness _____cm

3. Use the ruler to measure the length, width, and thickness of your lab notebook in centimeters. Record your results.

Length _____ cm   Width _____ cm   Thickness _____ cm

4. Calculate the volume of the lab notebook in cm^3 using both the observed and converted units.

Observed _____   Converted _____

POST-LAB QUESTIONS

1. How do you account for the difference between the volume calculated from converted units, and the volume calculated from the directly measured centimeters?

2. Which volume is more accurate?

3. Which measurements were more accurate, inches or centimeters? Why? What about the measuring equipment made a difference?

PROCEDURE 1.2: QUALITATIVE DENSITY

1. Into a large graduated cylinder, add carefully and slowly in this order:

    a. Corn syrup

    b. Water with food coloring

    c. Vegetable oil

d. Ethanol

2. Now, add a cork, ice cube, pebble, piece of chalk, paper clip, and aspirin, and record your observations. A picture is the best way to do this.

POST-LAB QUESTIONS

1. Sketch a density column showing the liquids and the position taken by each of the solids in the column.

2. Is solid water (ice) more or less dense than liquid water? How do you know?

3. Will the densities of different temperatures of the same liquid vary? Will hot water be more or less dense than cold water?

4. Why is it important to pour the column slowly? What will happen if it is shaken?

PROCEDURE 1.3: QUANTITATIVE DENSITY

Density of a Solid

1. Take a glass marble and a metal marble and measure the diameter of each using calipers, if available, or a centimeter ruler.

   Glass _____cm   Metal _____cm

2. Calculate the volume of each marble using the following formulas. Show your work.

   $r = \frac{1}{2}d$, $V = 4/3\pi r^3$, $\pi = 3.14$

   Glass _____cm^3 Metal _____cm^3

3. Determine the mass of each of the marbles.

   Glass _____g   Metal _____g

4. Fill a 25 mL graduated cylinder to about 10 mL with water. Record the precise volume, remembering to view the cylinder straight on, not from an angle, and to read from the bottom of the *meniscus* (the bowl-like shape adopted by the water).

   Volume =

5. Add the metal marble to the cylinder and again record the precise volume.

   Volume 2 =

6. Add the glass marble to the cylinder and again record the precise volume.

   Volume 3 =

7. Using the volumes recorded above, and remembering that 1 mL = 1 cm^3, calculate the displacement volume of the marbles. Show your work.

Glass _____ cm^3 Metal _____ cm^3

8. Using both the calculated volume and the displacement volume, determine the density of the glass and metal marbles. Show your work.

Calculated volume:

Glass _____ g/cm^3 Metal _____ g/cm^3

Displacement volume:

Glass _____ g/cm^3 Metal _____ g/cm^3

9. Obtain a density unknown and, using either measurement and calculation of volume or the displacement method, determine the density of the unknown sample and its identity. Show your work.

Unknown # _____

Mass:

Volume:

Density:

Identification:

POST-LAB QUESTIONS

1. How do you account for the difference between the volume calculated from measurement and the volume calculated from displacement? Which do you think is more accurate? Why?

2. If there is a small error in a measurement, what is the result of using that measurement in calculations?

3. What method did you use to determine the volume of your density unknown? Why did you choose this method?

4. What is the identity of your unknown?

Density of a Liquid

1. Fill a 100 mL beaker about half-full with room-temperature distilled water. Measure and record the temperature in °C.

   Temperature of distilled water _____ °C

2. Weigh an empty 50 mL beaker and record the mass.

   Mass of 50 mL beaker _____ g

3. Using a 10 mL volumetric pipet, obtain *exactly* 10 mL of the distilled water and transfer to the weighed beaker. Weigh the beaker again, and subtract to get the exact mass of the 10 mL of water.

   Mass of 50 mL beaker + 10 mL distilled water _____ g

   Mass of water alone _____ g

4. Calculate the density of water at the temperature you recorded. Repeat the procedure to check your precision. To check your accuracy, compare to the standard values provided here.

   Density of water at ____ °C = _____ g/mL

Repeat density of water at _____ °C = _____ g/mL

Standard densities of water at various temperatures. From CRC Handbook, 88th Edition	
20.0 °C	0.9982063
20.5 °C	0.9981019
21.0 °C	0.9979948
21.5 °C	0.9978852
22.0 °C	0.9977730
22.5 °C	0.9976584
23.0 °C	0.9975412
23.5 °C	0.9974215
24.0 °C	0.9972994
24.5 °C	0.9971749

5. Repeat the same procedure with the three colas and the isopropyl alcohol.

Mass of 10 mL regular cola _____ g

Density of 10 mL regular cola _____ g/mL

Mass of 10 mL diet cola _____ g

Density of 10 mL diet cola _____ g/mL

Mass of 10 mL lemon-lime cola _____ g

Density of 10 mL lemon-lime cola _____ g/mL

Mass of 10 mL isopropyl alcohol _____ g

Density of 10 mL isopropyl alcohol _____ g/mL

POST-LAB QUESTIONS

1. Arrange the five liquids—water, isopropyl alcohol, and the three colas—in order from least dense to most dense. If there are any that cannot reasonably be distinguished, identify them.

a. Least dense =

b.

c.

d.

e. Most dense =

2. Of the colas, what causes a difference in density? Can you identify the ingredient that causes the density to be higher in one type of cola than another?

3. Why is it important to use a volumetric pipet in these experiments rather than just pouring up to the 10 mL line on the beaker?

Density of a Gas

1. Get a 250 mL flask and a balloon. Weigh the two together and record the weight.

   Mass of flask + balloon _____ g

2. Place a small piece of dry ice (solid carbon dioxide) in the flask, quickly seal the flask with the balloon and record the total mass.

   Mass of flask + balloon + dry ice _____ g

3. Allow the carbon dioxide to sublime, and fill the flask and balloon (feel free to complete another section while you wait, or put the flask into warm water to speed up the process), then measure the circumference of the balloon. Calculate the volume of the gas using the following formulas:

   circumference = $\pi \cdot$ diameter(d), $\pi = 3.14$

   radius (r) = 1/2 d

volume $= 4/3\pi r^3$

Using this data, and the mass of the gas, calculate its density in g/L.

Circumference = _____ cm

Radius = _____ cm

Volume of balloon = _____ cm^3

Mass of carbon dioxide = _____ g

Density of carbon dioxide = _____ g/mL

4. Tie off the balloon, and observe its behavior. Does it float in the air?

POST-LAB QUESTIONS

1. What is the density of carbon dioxide gas? Is it higher or lower than the density of solid carbon dioxide?

2. Is carbon dioxide gas more or less dense than air? Why?

3. When determining the volume of the gas, why is it not necessary to add in the volume of the flask?

PROCEDURE 1.4: SPECIFIC HEAT

1. Use one of the specific heat samples to determine the minimum quantity of water required to *completely* cover the metal sample (disregard the hook, if there is one) in a small beaker. Remove the metal sample, and measure the quantity of water by pouring it into a graduated cylinder.

   Volume of water required _____ mL

2. Dry the beaker, place it on the balance, and *tare* the balance (set it to zero). Pour in the exact quantity of water recorded above, and record the mass.

   Mass of water required _____ g

3. Record the temperature of the water.

   Temperature of water _____ °C

4. Take the beaker, thermometer, and a pair of tongs over to the beaker where the other specific heat samples are in boiling water. As quickly as possible, move one of the samples from the boiling water to your beaker. Measure the temperature until it stops changing. This will happen quickly.

   Final temperature of water _____ °C

5. Carefully dry off the unknown (its temperature will be the same as that of the water), and determine its mass.

   Mass of unknown metal _____ g

6. Use the following steps to determine the specific heat of the unknown.
   a. Calculate the heat absorbed by the water.

   Heat = mass of water x specific heat of water x temperature change (final – initial)

   Heat = _____ g x 1.0 cal/g • °C x (_____ °C – _____ °C)

Heat absorbed by water = _____ calories

b. This same quantity of heat was lost by the metal. As the heat left the metal, it is given a negative sign. Use the following formula to calculate the specific heat of the metal.

(Heat absorbed by water) = mass of metal x specific heat of metal x temperature change (final temp. – 100 °C)

Specific heat of metal _____ cal/g • °C

Identity of metal _____

POST-LAB QUESTIONS

1. How do you know that the initial temperature of the metal is 100 °C?

2. Why was the change in the metal temperature so dramatic and the change in the water temperature so small?

3. Why did we want to use the minimum quantity of water?

Chapter 2

Atomic Structure and Nuclear Radiation

The word "atom" was coined by the Greek philosopher Democritus around the year 450 B.C. He was systematizing an idea, developed even earlier, that at some point it will no longer be possible to divide matter—you will reach an "ατομος" or "uncuttable" state. It was a long road of discovery from this philosophical idea to modern atomic theory.

In 1789, Antoine Lavoisier developed the law of conservation of mass. He also defined elements as substances that could not be broken down any further by chemical means. John Dalton's observation that elements always react in predictable ratios led to his development of his atomic theory. Proposed in 1805, Dalton's theory contained Democritus's idea that atoms cannot be subdivided, Lavoisier's idea that elements can't be broken down any further, and the additional idea that atoms of a given element are identical and share properties. He further proposed that what is happening in chemical reactions is the rearrangement of atoms. It was not until 1897 that a model of the atom was proposed by J. J. Thompson. After Thompson's discovery of the electron, he further proposed that atoms must also contain a positive charge, so he hypothesized that the electrons are embedded in a sea of positive charge, sort of like raisins in a pudding.

In 1900 physics and chemistry collided when Max Planck demonstrated that atoms can only emit energy in discrete ways. In 1905 Einstein explained that absorbing discrete packets of energy can dislodge electrons from atoms. In 1911 Ernest Rutherford performed an experiment giving us the model of the electron used to this day. He determined that the positive charge was concentrated in the center of the atom, while the electrons, with their negative charge, were in orbit around this center.

This partly satisfied the observations of scientists regarding the behavior of atoms, but left one thing unexplained. In one of the experiments in this chapter, you will demonstrate that particular elements emit particular colors when energy is added to them. This is because the color is distinctive enough that it can be used to identify the element.

Niels Bohr, in 1913, built on the work of Planck, Einstein, and Rutherford by proposing that electrons can exist only in discrete orbits. They can travel between orbits but only by jumping: they exist in their orbits, but never between them. It is an unsatisfying idea because in the world around us we do not see anything else that behaves this way.

These ideas led to a new branch of physics—quantum mechanics. Quantum mechanics, however, is a field of study for the super-small, microscopic world. Nothing in the world of quantum mechanics is large enough to see with the naked eye.

So, is it all figured out? Remember our friends at CERN? Far from being the indivisible marbles of Democritus and Dalton, the more closely modern scientists "look" at atoms, the more wonderfully complicated they discover they are.

1. The half-life of actinium-225 is 10 days. If you begin with 10 g of actinium-225, how much will be left after:

    a. 10 days?

    b. 20 days?

    c. 30 days?

    d. 60 days?

2. Classify the following as metals or nonmetals.

    a. Actinium

    b. Phosphorus

    c. Calcium

    d. Uranium

3. In procedure 2.2, why do different metals give off different colors if heated in a flame?

MATERIALS

100 pennies per student

Shallow boxes large enough to hold all the pennies in one layer

Nichrome wire loops (along with hydrochloric acid for cleaning) or wooden splints

Bunsen or alcohol burners

Lithium chloride

Strontium chloride

Potassium chloride

Barium chloride

Zinc chloride

Copper chloride

Spectroscopes

Two graphite pencils per lab group, sharpened on both ends

Strip of cardboard for each lab group

One 9V battery per lab group

Copper wire and wire cutters

200 mL beaker per lab group

Water

Sodium chloride

Steel or glass marbles (25 or more of two colors per lab group)

Plastic marbles

Balance or postal scale

PROCEDURE 2.1: HALF-LIFE

1. Put 100 pennies heads-up in the bottom of a box.

2. Give the box a quick shake, and remove and count the pennies that are now tails-up. Fill in the chart of shakes-versus-heads remaining:

Shakes	Heads
0	100
1	
2	
3	
4	
5	
6	
7	
8	
9	
10	

3. Graph the data using the grid below. Be sure to label the axes. Plot shakes on the *x*-axis and heads on the *y*-axis.

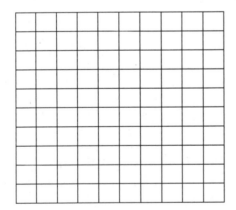

POST-LAB QUESTIONS

1. Is the graph a straight line? Why or why not?

2. A gram of radioactive uranium (U-235) contains about 2.5×10^{21} atoms. If it has a half-life of around 69 years, how many radioactive atoms will remain after 276 years?

3. Radioisotopes are sometimes referred to as "unstable." Given the above example, provide a working definition of nuclear instability.

PROCEDURE 2.2: QUALITATIVE ATOMIC IDENTIFICATION

1. If you are using nichrome wire loops, clean them carefully by dipping them in hydrochloric acid. Light the burners provided, and burn off the acid. No color should be seen.

2. Dip the wire or splint into the solution or salt provided. If using splints and salts, first dampen the splint slightly so the salt will stick.

3. Place the wire or splint in the flame of the burner and record the color you see.

4. Repeat for all the salts provided, as well as the unknown. If using a wire, clean it in acid before each use.

5. If spectroscopes are available, examine the flame alone, then look at each flame test through the spectroscope. Note the number and color of the lines in each.

POST-LAB QUESTIONS

1. How do you account for the different colors seen?

2. Each of the salts is a metal bonded to chlorine. Which accounts for the color? How do you know this?

3. What is the identity of the unknown?

4. How many lines were seen with each element? How does this correspond to the molecular weight of the element? Do larger elements have more or fewer lines? Why?

PROCEDURE 2.3: FROM COMPOUND TO ELEMENT

Atoms combine to form compounds. Although radioactive isotopes can decay into other atom types, combinations of atoms can be separated relatively simply.

1. Make sure that the pencils are sharpened on both ends.

2. Push the pencils through the cardboard about 1 inch apart. It is important that the pencils do not touch.

3. Attach the negative terminal of the battery to the graphite of one pencil using the copper wire.

4. Attach the positive terminal of the battery to the graphite of the other pencil on the same side of the cardboard.

5. Stir a pinch of salt into enough warm water to fill the beaker to within an inch or so of the lip.

6. Lower the ends of the pencils not connected to the wires into the salt water.

POST-LAB QUESTIONS

1. Record your observations. What happens when the electrified graphite is lowered into the salt water?

2. Knowing that the formula for water is H_2O, what do you expect that you are seeing?

PROCEDURE 2.4: BUILD-YOUR-OWN ISOTOPES

1. Take a collection of marbles. For the purposes of this exercise, the glass or metal marbles are your neutrons and protons, and the plastic marbles are your electrons.

2. First, decide which color of glass or steel marble is the proton and which is the neutron. Next, build neon-22. Include protons, neutrons, and electrons. Take the container of marbles and weigh it.

3. Build and weigh neon-20 and neon-21, and record their weights.

POST-LAB QUESTION

1. In nature, the relative abundance of these isotopes is 90.48% for neon-20, 0.27% for neon-21, and 9.25% for neon-22. Using these abundances and the weights of your "isotopes," calculate an average "atomic weight" for your neon atoms. Be sure to show your work.

Chapter 3

Compounds and Molecules

The tendency of atoms to combine with one another does not simply lead to a variety of differently inert combinations like very small molecular models. The different properties of atoms cause them to react in different ways, thus creating compounds that form to have different physical and chemical properties.

Atoms have very different abilities to capture and hold electrons, and this will create different types of compounds. If one atom is very good at holding on to electrons, as nonmetals tend to be, and another has a hard time holding on to electrons, as metals tend to be (which is why they conduct electricity), they will not be able to form a bond that requires electrons to be shared. Instead, one or more electrons will be transferred from one atom to the other. The metal will give up an electron or two, and the nonmetal will gain an electron or two, causing both to become *ions*. This does not really form a proper bond, though the attraction of the resulting two charges can be very strong. Anyone who has ever taken a fall on an icy sidewalk covered with rock salt can attest to the exceptional hardness and complete lack of give exhibited by rock salt crystals bound together by opposite charges.

What is fascinating about these compounds, bound together by ionic bonds, is how exceptionally tightly they hold their electrons and how tightly they hold together. Electrons do not flow through them, their melting points approach 1000 °C, and a salt crystal is hard to break even with a hammer. When it does break, it tends to shatter. But when you drop that same crystal into water, a polar liquid, it comes apart without needing any addition of energy and is soon completely dissolved. Ionic compounds are usually perfectly content to substitute the single full charge of the opposite ion for the multiple small charges offered by the polar liquid.

So what creates a polar liquid? There are atoms that have similar affinities for their electrons but could use another electron or two to maximize their stability. With a small addition of energy, these atoms will share electrons. Their physical and chemical properties will depend largely on how *well* they share. Atoms that are identical share electrons perfectly. The electrons spend exactly the same amount of time attached to each of the atoms, and there is no distribution of charges. These molecules are covalently bonded to one another because they are sharing electrons, and they are nonpolar. They will not dissolve salts, and they will have a hard time mixing with the molecules that do dissolve salts. The polar molecules, on the other hand, have trouble sharing the electrons: one of the atoms tends to "hog" the electrons making that atom slightly negative, while the other atom has trouble hanging on to its electrons, making it slightly positive.

So, here's where it gets interesting: polar molecules will only dissolve in polar liquids, while nonpolar molecules will only dissolve in nonpolar liquids. This property makes it possible to identify molecules and liquids by their polarity.

PRE-LAB EXERCISE

1. Identify the following as nonpolar or polar. Draw Lewis structures to support your answer.

   a. Calcium chloride

   b. Carbon dioxide

   c. Formaldehyde (CH_2O)

   d. Potassium iodide

2. Classify the following as metals or nonmetals.

   a. Actinium

   b. Phosphorus

   c. Calcium

   d. Uranium

3. How many valence electrons does each of the following have?

   a. Hydrogen

   b. Nitrogen

   c. Oxygen

   d. Carbon

   e. Chlorine

   f. Sodium

   g. Magnesium

MATERIALS

Conductivity testers

Distilled water

Sodium chloride

Potassium chloride

Sucrose

Fructose

Water

Isopropyl alcohol

Magnetic stirrer and stirring bars

100 mL beakers, or other size to accommodate conductivity testers

Hexanes or light oil

Chromatography strips

Acetone

Food coloring

PROCEDURE 3.1: IONIC OR COVALENT?

1. Fill a 100 mL beaker halfway with distilled water. Put it on a stir plate.

2. Put the conductivity tester in the water, and record your observations.

3. Dissolve a large pinch of one of the compounds in the water, and lower in the conductivity tester. Record your observations.

4. Repeat for the other compounds, then for the unknown, and be sure to scrupulously clean the beaker between each measurement to avoid cross-contamination. Make a table, and record whether each compound, including distilled water and the unknown, if there is one, is polar or nonpolar.

POST-LAB QUESTIONS

1. Given your observations with distilled water, why is there an electrocution hazard for swimmers in a thunderstorm?

2. What type of bonding is associated with compounds that dissolve and conduct electricity?

3. What sort of bonding is associated with the compounds that dissolve but do not conduct electricity?

4. Are any of the compounds nonpolar? Why or why not?

PROCEDURE 3.2: POLAR OR NONPOLAR?

1. Carefully mix the following substances to see what the results are. Do the solids dissolve? Are the liquids miscible?

Solvent	Solute	Soluble? (solids)	Miscible? (liquids)
Water	Sodium chloride		
Water	Sucrose		
Water	Hexanes		
Water	Isopropyl alcohol		
Isopropyl alcohol	Sodium chloride		
Isopropyl alcohol	Sucrose		
Isopropyl alcohol	Hexanes		
Hexanes	Sodium chloride		
Hexanes	Sucrose		

POST-LAB QUESTIONS

1. Which of the mixtures do you expect to conduct electricity? Why?

2. What types of bonding are seen in the compounds used? Use your knowledge of the structures and/or your observations of their behavior to justify your answers.

   a. Water

   b. Isopropyl alcohol

   c. Hexanes

   d. Sodium chloride

   e. Sucrose

3. In the mixtures in which the substances mixed completely, what intermolecular forces were responsible for this mixing? (A case not mentioned in the book but that may be present here is an ion-dipole interaction.)

    a. Mixture 1 =

    b. Mixture 2 =

    c. Mixture 3 =

    d. Mixture 4 =

    e. Mixture 5 =

    f. Mixture 6 =

4. For liquids that did not mix, which formed the top layer, and which formed the bottom layer? Which liquid was the most dense? The least dense? Why?

PROCEDURE 3.3: MODEL MOLECULES

Atoms combine to form compounds. Draw Lewis structures of the following molecules, and build models.

1. H_2

2. O_2

3. N_2

4. H_2O

5. H_2O_2

6. NH_3

7. CH_4

8. CH_2O

9. CO_2

10. C_2H_6

11. C_2H_4

12. C_2H_2

13. HCN

14. HOCN

POST-LAB QUESTIONS

1. Which of the atoms in these molecules can form only single bonds?

2. Which of the atoms in these molecules can form only single or double bonds?

3. Which of the atoms in these molecules can form single, double, or triple bonds

4. Based on your answer to the previous question, use the periodic table to name another atom that can form: only single bonds; single or double bonds; or single, double, or triple bonds?

5. A part of Dalton's atomic theory not mentioned in your book is his "rule of greatest simplicity" which states that when atoms combine in a ratio, it must be assumed to be binary (one of each). Was this correct? Why or why not?

PROCEDURE 3.4: MOLECULAR SEPARATIONS: CHROMATOGRAPY

1. Place a small quantity of water in a beaker and a small quantity of acetone in another beaker.

2. Cut two strips of chromatography paper to the specifications provided by your instructor.

3. Place a drop of ink or food coloring ½ inch from one end of each strip.

4. Put one strip into the beaker of water and the other strip into the beaker of acetone, with the ink side closest to the solvent.

5. Allow the solvent to travel up the paper by capillary action, and observe the effect that the different solvents have on the colored solution.

6. Allow the paper to dry.

POST-LAB QUESTIONS

1. If a colored solution is separated more effectively by water than by acetone, what does that tell you about the nature of the colored compounds in the solution?

2. What can you say about the molecular weight of the compounds that travel the farthest? The least far?

3. Are the colored solutions mixtures of different molecules, or all the same molecules?

Chapter 4

Chemical Quantities and Chemical Reactions

Although chemical reactions occur all around us all the time, we are rarely aware of them. All in all, at least to outward appearances, our world is pretty stable. We are used to the amazing reactions taking place all around us, so they don't come as any surprise. Chemical reactions surround us when we convert oxygen to carbon dioxide inside of our bodies, plants convert carbon dioxide to sugar through photosynthesis, and our cars burn gasoline or use electricity to get us to chemistry lab. All these reactions, commonplace and ordinary though they may seem, are what drive our world.

In biological systems, reactions, such as those that take carbon dioxide and water and create sugar in plants, do not happen on their own (though it is theoretically possible that they could). If they did happen spontaneously, they would happen too slowly to be really useful. Something usually has to be added to speed things up, and in the industrial world, catalysts are used. In biological systems, enzymes are used. Sometimes, as in one of the experiments in this chapter, a catalyst or an enzyme can do the same work.

So, what's the difference between a catalyst and an enzyme? First, let's consider what is the same. Both enzymes and catalysts speed up reactions by reducing the energy required to make the reaction happen. They are not affected or consumed by a reaction, and they can be used repeatedly. The primary difference is in their origins: a catalyst typically does not have a biological origin but an enzyme does. Enzymes are very high molecular weight proteins that are fine tuned to operate best under the conditions in which they expect to find themselves in a biological system. So, for instance, there are digestive enzymes that work best in very acidic conditions, such as those found in the stomach, but there are also digestive enzymes that work best in the more neutral conditions provided by saliva and the small intestine. All digestive enzymes, though, work best at human body temperature, 98.6 °F.

When reactions occur, the proportions of reactants to products will always be the same whether or not there was an enzyme or catalyst involved. To use a common kitchen example, if you are making sandwiches, you will get 1 sandwich from every 2 slices of bread:

$$2 \text{ Bread} \xrightarrow{\text{your work}} 1 \text{ Sandwich}$$

Silly though this example is, it makes its point—the proportion of sandwiches to bread will be the same regardless of the amount of bread you started with. If you are going to get fancy making club sandwiches, for instance, your proportions will be different but the reaction will be, too. In nature, there are equivalents: individual oxygen atoms normally join together to form oxygen atoms, O_2. Under fancy conditions, lightning storms, for instance, oxygen can get fancy and make club sandwiches, too— in the form of ozone, O_3. If you know the proportions and the conditions, you can predict the quantities of reactants or products.

1. Write and balance the equation for the decomposition of hydrogen peroxide (H_2O_2).

2. Write and balance the equation for the oxidation/reduction reaction of zinc and hydrochloric acid, which produces hydrogen gas and Zn^{2+} ions.

3. Write and balance the equation for the reaction of baking soda ($NaHCO_3$) and acetic acid (CH_3COOH), the active ingredient in household vinegar. This reaction generates sodium acetate ($NaCH_3COO$), carbon dioxide, and water.

4. Calculate the molecular weight of baking soda.

5. Calculate the molecular weight of vinegar.

MATERIALS

3% H_2O_2

Frozen liver, cut into small pieces

Manganese dioxide

Metal paper clip

Lab timer

Balance or postal scale

Pennies

Metal snips or hacksaw

1 M or 1 N hydrochloric acid

Baking soda (dry completely using a desiccator or drying oven, if possible)

Household vinegar

Magnetic stirrers and stirring rods

PROCEDURE 4.1: ENZYMES AND CATALYSIS

1. Fill four 50 mL beakers halfway with 3% hydrogen peroxide.

2. Label one of the beakers "control," one "liver," one "MnO_2," and one "paper clip."

3. Get the timer ready, and place a tiny piece (a few cubic millimeters) of liver into the labeled beaker. Record the amount of time it takes for the reaction to stop.

4. Place a small pinch of manganese dioxide into the other labeled beaker. Record the amount of time it takes for the reaction to stop.

POST-LAB QUESTIONS

1. Given the nature of the substances you added to the hydrogen peroxide, what can you deduce about what is causing the response of the solution?

2. If you used 50 g of hydrogen peroxide, how many grams of oxygen gas have been generated by the reaction?

3. How many grams of oxygen per second does this reaction generate before the hydrogen peroxide is used up?

4. Roughly how much oxygen was generated by the degeneration of the hydrogen peroxide in the other beaker in the same amount of time?

5. If the rate is constant, how long would it take for 100 g of hydrogen peroxide to react?

6. How many grams of oxygen would be generated by the reaction of 100 g of hydrogen peroxide?

7. Which is better at speeding up the reaction, the liver, the paper clip, or the manganese dioxide? Rank them in order of catalytic ability.

PROCEDURE 4.2: STOICHIOMETRY AND MASS PERCENT

1. Weigh a post-1982 penny. Assume that the entire penny is made of zinc. Write down the mass in grams.

2. Using a balanced equation, and assuming that the entire mass of the penny is zinc, calculate the number of moles of hydrochloric acid needed to completely dissolve the penny.

3. If the acid available has 1 mol of hydrochloric acid per 1000 mL, how many milliliters of acid will be needed to completely dissolve the penny?

4. Put approximately 150% of the necessary amount of acid in a beaker, cut the penny to uncover the zinc (copper does not react with hydrochloric acid), and place the penny in the acid.

5. Allow the reaction to go to completion (there will be no more bubbles when this is the case). This may take a while, but it can be speeded up with a hotplate or stirrer. Rinse off the remaining copper shell gently with water, and then with isopropanol to speed the drying.

6. Weigh the remaining copper.

POST-LAB QUESTIONS

1. Write down the total mass of the penny.

2. If the copper shell weighs 0.0625 g, how many grams of zinc are present in the

penny?

3. How many moles of zinc are present?

4. How many moles of copper are present?

5. What is the mass percent of copper and zinc in the penny?

6. What could be done to speed the reaction?

7. If acid with 5 moles of acid per 1000 mL was available, how much would be needed to dissolve the penny?

PROCEDURE 4.3: HOW MUCH ACID?

1. Carefully measure out exactly 50 mL of vinegar using a graduated cylinder, and pour it into a 250 mL beaker. Put the beaker on a stirplate.

2. Weigh a weigh boat or piece of weighing paper, then, without zeroing the balance, add approximately 5 g of dried baking soda. Write down the exact weights here.

   a. Weighing paper or weigh boat:

   b. Baking soda:

3. Set the stirplate stirring, and slowly add the baking soda a little at a time, pausing in between small additions until the reaction stops. As soon as the next small addition causes no more reaction (you will be able to see the reaction slowing, so you can adjust the size of the additions), weigh the remaining baking soda and write the weight here.

4. Calculate the number of grams of baking soda used to neutralize the vinegar.

5. Using the molecular weight calculated above, how many moles of baking soda did you use?

6. Using the balanced equation above, how many moles of acetic acid were there in the 50 mL of acid?

POST-LAB QUESTIONS

1. What is the concentration of vinegar in moles per mL?

2. Using the molecular weight of acetic acid from the pre-lab, how many grams of acetic acid were in the 50 mL of vinegar?

3. What is the percent concentration of vinegar, if the 50 mL of vinegar weigh 50 g?

4. A large box of baking soda is often kept in the lab or kitchen. Aside from its ability to neutralize acid spills, it is also used for fighting fires. Write an equation for the decomposition of baking soda, then explain why this compound is especially effective for extinguishing fires.

Chapter 5

Changes of State and the Gas Laws

Our lives are water-based, and it is not uncommon for us to be in the presence of all three phases of this remarkable liquid. Whether sitting outside on a humid day with a glass of iced tea, or engaging in a polar bear swim in icy conditions, it is not unusual for us to be around water in its solid, liquid, and gaseous forms. When it changes phases, this molecule behaves like almost no other. When boiled on a stove, it behaves as expected: a small quantity of liquid generates a large amount of gaseous steam. But when it is frozen, it does something remarkable. In contrast to just about everything else in our experience, when water freezes it does not shrink, but expands. This is something we have discovered if we have ever put a can of soda into the freezer "just for a minute" or tried to freeze a glass jar of soup without leaving enough head space: the liquid expands strongly enough to break the can or shatter the glass. Something just as unique reminds us that solid water is special, something that we don't often notice because we have seen it over and over again our whole lives: ice floats.

If ice did not float, it seems unlikely that life would exist on Earth in the abundance with which it does. Every winter, a thin layer of ice would form and sink, submerging the cold ice under the insulating water. When summer came it would take a very long time for the water to warm up enough to melt even some of the submerged ice, and breezes blowing across the icy water would keep the surroundings cold as well. Only the hardiest creatures could live in those waters or on their banks. Instead, ice forms on the surfaces of aquatic bodies and stays there without sinking, allowing the creatures in the water below to continue living under the ice all winter. When the weather warms, the surface ice is exposed immediately to the increased warmth and melts. The insulating power of water modulates the weather around it, while supporting all manner of plant and animal life. What a remarkable set of unique features for something we take so much for granted.

We live in a sea of gases, but almost never do we stop and appreciate that interesting fact. Our bodies require the oxygen in the air and generate carbon dioxide that we then exhale. When we dive under the water, we become very aware of the weight of water pressing down on us. But when we are out of the water, we don't even notice the fact that we are carrying around nearly a ton of air at every moment. Unlike liquids and solids, the density of gases is highly temperature-dependent. In the atmosphere, temperature variations (along with humidity and other variables) cause the density of the air to vary. Naturally, air will move from areas of higher density to lower density, one of the primary causes of surface wind. Clouds in one area will cause the air to cool, making that air more dense, while an area close by experiencing lots of sunshine will have air that is less dense. When the sun sets, this process stops, and surface winds tend to die down.

1. Write and balance the equation for the reaction of baking soda ($NaHCO_3$) and acetic acid (CH_3COOH), the active ingredient in household vinegar. This reaction generates sodium acetate ($NaCH_3COO$), carbon dioxide, and water.

2. If 10 g of baking soda were added to an excess of vinegar, how many moles of carbon dioxide would be generated?

3. If the temperature in the lab is 22 °C, what volume of gas would be produced, assuming that CO_2 is an ideal gas?

4. Using the equation, $V = 4/3\pi r3$, what would be the radius of a sphere that contained this volume of gas?

5. Using the equation $C = 2\pi r$, what would be the circumference of a sphere that contained this volume of gas?

MATERIALS

Mossy zinc or zinc strips

5 M hydrochloric acid

Scale or balance

100 mL Erlenmeyer flask

Balloons

Thermometer

Flexible measuring tape (cm)

Gloves

Forceps

Dry ice (solid CO_2)

250 mL graduated cylinder

Hotplate

PROCEDURE 5.1: CALCULATING THE VOLUME OF A GAS

1. Carefully place about 50 mL of hydrochloric acid in a 100 mL Erlenmeyer flask and obtain a small piece of zinc. Use a gloved hand or forceps to carry the zinc, not because it is dangerous or toxic but because the oils on your skin can impede the reaction we wish to observe.

2. Write down the mass of zinc in grams.

3. Convert the mass of zinc to moles.

4. Using the fact that zinc and chloride react to form a salt with the formula $ZnCl_2$, balance the equation for the reaction between zinc and hydrochloric acid. Other than the salt, what will be produced?

5. With this balanced equation and the number of moles of zinc you have, calculate the moles of gas that will be produced, keeping in mind that this is a diatomic gas.

6. If the molar volume of a gas is 22.4 L at 0 °C, use Charles's law to calculate the molar volume of a gas at the temperature of the lab.

7. If the zinc reacts completely with the acid, what volume of gas will be generated? If the amount is more than 2 L, use a smaller piece.

8. Now, test your calculation. Place the zinc inside the balloon, and put the balloon over the neck of the flask. Drop the zinc from the balloon into the acid, and allow it to completely react. When the solution is no longer bubbling, the reaction is complete.

9. Assume that the balloon is perfectly round (which is, of course, incorrect as the balloon is *not* perfectly round) and measure the circumference in centimeters. Using the equations below, find the approximate volume of the balloon in cubic centimeters.

$$C = 2\pi r$$
$$V = (4/3)\pi r^3$$

10. If your instructor is very brave, he or she will allow you to tie off the balloon and touch it to a flame to see another reaction.

$$2\,H_2 + O_2 \rightarrow 2\,H_2O$$

11. If you will be completing Procedure 5.2, carefully tie off the balloon.

POST-LAB QUESTIONS

1. Convert the volume from cubic centimeters (milliliters) to liters.

2. How close was your estimate? Provide a percentage and how you calculated it.

3. In almost every reaction, one reagent runs out first. This is the limiting reagent. Look at the reaction vessel: what do the remaining contents tell you about which was the limiting reagent?

4. When the balloon is placed on the neck of the flask, the flask is full of air. Does the volume of the air need to be taken into account? Why or why not?

PROCEDURE 5.2: PHASE CHANGES AND VOLUME

1. Place a 100 mL flask on the balance and tare it (set it to zero). With the balloon in hand, put a small piece of dry ice in the flask and *quickly* note the weight. If it is more than 4 grams, take the flask off the scale for a bit to allow some of the carbon dioxide to sublime, them put it back on. Note the exact weight, but do not leave the dry ice on the balance for very long. Also, write down a rough estimate of the volume of the dry ice in cubic centimeters.

2. Immediately after weighing, put the balloon over the neck of the flask and allow the dry ice to sublime until it has completely disappeared.

3. While waiting for the gas in the balloon to equilibrate at room temperature, calculate the number of moles of carbon dioxide that were placed in the flask.

4. Using the molar volume of an ideal gas (22.4 L at 0 °C), calculate the molar volume of an ideal gas at the temperature of the laboratory using Charles's law.

5. Measure the circumference of the balloon, and use the equations above to determine the volume in liters.

POST-LAB QUESTIONS

1. Approximately how much did the volume of the carbon dioxide change when it sublimed?

2. Does the carbon dioxide come close to the molar volume of an ideal gas? Why or why not?

3. What accounts for the "depressed" behavior of the carbon dioxide-filled balloon compared to a regular balloon? How does it compare to the hydrogen-filled balloon?

4. The Hindenburg disaster was caused by a spark in the hydrogen-filled airship. What would be the result of exposing this balloon to a spark? Would this gas be useful for filling an airship? Why or why not?

PROCEDURE 5.3: TEMPERATURE CHANGES AND VOLUME

1. Carefully fill a 100 mL Erlenmeyer flask all the way to the brim. Invert the 250 mL graduated cylinder over the top of the full flask and carefully and quickly flip both over so that the water goes into the cylinder as completely as possible. Record the total volume of the flask here.

2. If it is available, swirl a little bit of acetone or ethanol in the flask to speed its drying.

3. Making sure that the balloon contains as little air as possible, place the balloon over the neck of a 100 mL Erlenmeyer flask at room temperature.

4. Using tongs, place the flask into a cooler where the dry ice is being kept and close the lid. After a minute or two, open the cooler, and remove the flask and balloon. Record your observations.

5. Still using the tongs, put the flask on a hotplate.

6. Once the balloon has inflated (use your best judgment), carefully measure the circumference, and remove it from the hotplate.

POST-LAB QUESTIONS

1. Using the same equations provided above, calculate the volume of the balloon.

2. Using the known temperature of the lab and the known volume of the gas in the flask, calculate the temperature of the air inside the balloon.

Chapter 6

Organic Chemistry: Hydrocarbons

Hydrocarbons, with a few heteroatoms thrown in here and there, are the basic building blocks on which almost the entire natural world is built. Twenty eight percent of the mass of your body is carbon and hydrogen. If you throw in oxygen, nitrogen, and phosphorus, the most common heteroatoms, you are over 97 percent. Although overall we are mostly water, carbon is an integral part of what makes our bodies work.

What makes carbon so amazing and interesting? Why would a writer like Isaac Asimov devote an entire book, aptly named *The World of Carbon*, to this element? Carbon is not unique in being able to bond to itself, but it is unique in being able to form long strings of atoms, called *catenation*. Other elements can do similar things, but they do them more as party tricks: the eight-atom rings of sulfur and the tetrahedral of phosphorus are both interesting and beautiful, but no one has ever discovered a sulfur-based life form.

When carbon forms long chains, or lattices, the resulting product has a remarkable stability because of carbon's ability to delocalize electrons. We talked about covalent bonds as electron-sharing arrangements, but no element does this quite as well as carbon, which will share not only with its neighbor, but also with atoms all up and down the chains. In Procedure 2.3, you took advantage of this in a way that you probably thought nothing of when you used double-sharpened pencils as electrodes. In earlier chapters, you undoubtedly learned that metals conduct electricity while nonmetals do not, unless the nonmetal is carbon in the form of graphite.

Even if limited only to binding with itself, carbon finds plenty of ways to be interesting. It occurs in numerous different forms, called allotropes, and in even more forms in the chemistry lab. Pencil leads are pure carbon. So are diamonds. So is coal. Graphene, a one-atom thick honeycomb lattice of carbon atoms, can be produced, albeit very carefully, using a pencil and a piece of tape. In the lab, things can get fun: practical applications may follow, but who can deny the inherent appeal of a buckyball, a molecule with the formula C_{60} that looks like a geodesic dome of the kind once popularized by the architect Buckminster Fuller?

Tubes (nanotubes) made of carbon lattices, not unlike buckyballs, show great promise as highly efficient conductors of electricity, and they may serve to drive the next wave of miniaturization in electronics. Super-sensitive speakers made of graphene sheets are only two atoms thick and perfectly clear. Foams made of carbon are about twice the density of air and are attracted to magnets, something scientists are still trying to get a handle on. At low temperatures, they can even act as magnets themselves.

So, common, everyday carbon is on the very cutting edge of science and physics: while burning coal powers electricity production, nanotubes carry that electricity where it needs to go.

PRE-LAB EXERCISE

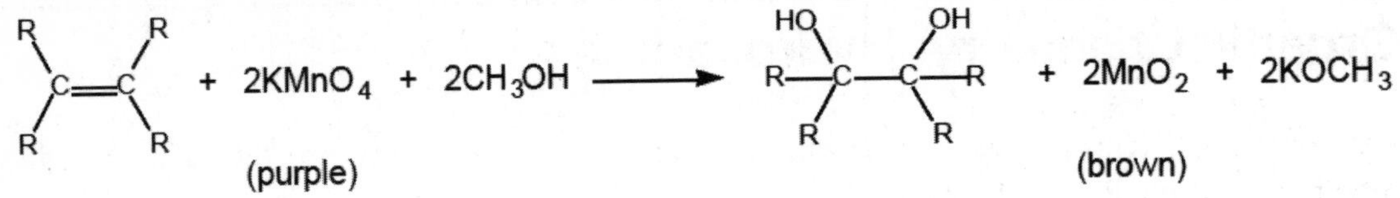

R
R
C
C
R
R
+ 2KMnO4 + 2CH3OH
(purple)
HO
OH
R
C
C
R
R
R
+ 2MnO2 + 2KOCH3
(brown)

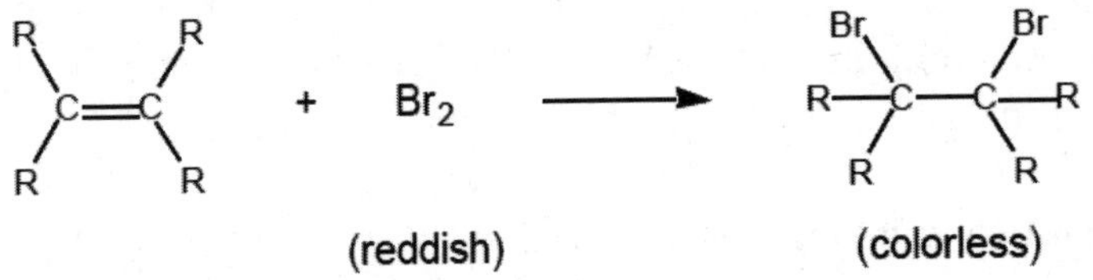

R
R
C
C
R
R
+
Br2
(reddish)
Br
Br
R
C
C
R
R
R
(colorless)

1. Write the formulas and draw the structures of the following molecules. Identify them as alkanes, alkenes, alkynes, or other, and predict whether or not they will react with potassium permanganate (top reaction) and/or bromine.

    a. Cyclohexane

    b. Cyclohexene

    c. Toluene

    d. Ethanol

    e. Dichloromethane

2. Which of the above molecules do you expect to be soluble in water? Why?

3. Which of the above molecules do you expect to be soluble in dichloromethane? Why?

MATERIALS

Gloves

Cyclohexane

Cyclohexene

Toluene

Dichloromethane

Bromine solution (very dilute solution in dichloromethane)

Potassium permanganate solution (2% $KMnO_4$ in water)

Ethanol

Test tubes

Test tube rack

Vegetable oil

Coconut oil

Lard

PROCEDURE 6.1: CHEMICAL PROPERTIES OF HYDROCARBONS

1. Label three test tubes "cyclohexane," "cyclohexene," and "toluene."

2. Put around 10 drops each of cyclohexane, cyclohexene, and toluene in the separate test tubes. Add 2 drops of dilute bromine in methylene chloride to each test tube.

3. Observe whether a reaction occurs (HINT: If the reddish color disappears, a reaction has occurred.)

4. If no reaction occurs immediately, put the tubes aside to see if a slow reaction occurs.

5. Label three more test tubes "cyclohexane," "cyclohexene," and "toluene." Dissolve 6 drops of each hydrocarbon in 40 drops of "ethanol" into each test tube. Label another tube ethanol, and place 40 drops of just ethanol into the tube.

6. Add 2 drops of $KMnO_4$ solution to each test tube. Observe any reaction.

7. Put the tubes aside to observe any slow reaction.

POST-LAB QUESTIONS

1. Did your results match the predictions you made in the pre-lab exercise? If so, explain why, and if not, suggest a reason.

2. What happened with the ethanol and potassium permanganate? Why was it important to include this tube?

3. Did any of the compounds exhibit a slow reaction with bromine? What might be happening in these slow reactions?

PROCEDURE 6.2: PHYSICAL PROPERTIES OF HYDROCARBONS

1. Label three more test tubes "cyclohexane," "cyclohexene," and "toluene." Add 10 drops of each compound to the labeled tubes. Carefully, add 20 drops of water to

each test tube and record your observations. Are the water and the organics miscible? If not, which is more dense? Less dense?

2. Shake the test tubes. If any of the combinations did not immediately mix, does this cause them to mix? Record your observations.

3. Repeat the test, using 10 drops of methylene chloride in place of the water. Note the miscibility and solubility of the compounds. Using a fourth test tube, check the miscibility and solubility of methylene chloride and water. In the case of any that are not miscible, which is more dense? Which is less dense? Record your observations in the chart below.

Solvent 1	Solvent 2	Miscible	More dense
Cyclohexane	Water		
Cyclohexene	Water		
Toluene	Water		
Cyclohexane	Methylene chloride		
Cyclohexene	Methylene chloride		
Toluene	Methylene chloride		
Water	Methylene chloride		

POST-LAB QUESTIONS

1. Which is a better predictor of density, molecular weight or intermolecular forces? Explain your answer.

PROCEDURE 6.3: NATURAL HYDROCARBONS AND THEIR PROPERTIES

1. Using a hot plate or burner, determine the relative melting points of the two solid fats. The oil is already a liquid at room temperature, so it has the lowest boiling point of the three substances.

2. Label two test tubes "vegetable," two tubes "coconut," and two tubes "lard" (for a total of six test tubes).

3. Put about 40 to 50 drops of cyclohexane in one of each of the oil test tubes, and 40 to 50 drops of water in the other test tube. You should now have two tubes for each natural hydrocarbon, with one containing water and the other one containing cyclohexane.

4. Add a few drops of vegetable oil to the vegetable oil tubes, and a small piece of coconut oil or lard to the others. Agitate to dissolve.

5. Test a tube of each with a few drops of bromine and potassium permanganate. Record the results in the below chart.

Compound	Solvent	Reagent	Soluble	Reaction
Vegetable oil	Cyclohexane	Bromine		
Coconut oil	Cyclohexane	Bromine		
Lard	Cyclohexane	Bromine		
Vegetable oil	Cyclohexane	$KMnO_4$		
Coconut oil	Cyclohexane	$KMnO_4$		
Lard	Cyclohexane	$KMnO_4$		
Vegetable oil	Water	Bromine		
Coconut oil	Water	Bromine		
Lard	Water	Bromine		
Vegetable oil	Water	$KMnO_4$		
Coconut oil	Water	$KMnO_4$		
Lard	Water	$KMnO_4$		

POST-LAB QUESTIONS

1. What were the relative melting points of the three natural hydrocarbons?

2. Which solvent did the compounds dissolve in? What does this tell you about the polarity of the compounds?

3. Which compounds reacted more readily with the bromine? With the potassium permanganate? Which solvent best facilitated the reaction?

4. What do the reactions tell you about the three compounds? Does this explain their different melting points? Why or why not?

Chapter 7

Organic Chemistry and Biomolecules

The number of small organic molecules is vast! As similar as they may seem, however, they can all be distinguished from one another with a number of chemical tests, which often need to be performed in series to distinguish an aldehyde from a ketone from an alcohol, etc. Even more remarkable, it is possible to distinguish a primary alcohol from a secondary alcohol from a tertiary alcohol.

This seems hardly possible until you consider the different ways these molecule types are formed. In all of them, there is an oxygen, one of nature's more reactive elements, located in such a way that makes it more or less accessible for further reactions, with either an electron-rich double bond, or two single bonds.

There's a reason that molecular models are standard equipment for chemistry students. It is hard to retain all this information and visualize what many of these terms mean. If you follow the discussion in class and lab, building models as you go, it will be time well spent.

Think of a simple aldehyde, R-HC=O. This is a pretty exposed location for the oxygen. An oxidizing agent will turn an aldehyde into a carboxylic acid with no trouble at all. A ketone, $R_2C=O$, similar though it is to the aldehyde, has a more protected oxygen. With no hydrogen next door, this compound will not convert into a carboxylic acid.

Some tests are just looking for a carbonyl group, C=O. These will react with both an aldehyde and a ketone, but are not at all reactive with alcohols.

Consider oxidizing agents that convert aldehydes into acids. They'll do the same thing for a primary alcohol, with two hydrogens nearby, or a secondary alcohol, with one hydrogen nearby. Tertiary alcohols have no hydrogen to oxidize, so these will not react with oxidizing agents.

Use the information below to work the pre-lab exercise, interpret your results, and design your experiments!

Organic	Jones reagent	Lucas reagent	2,4-DNP
Primary alcohol	Blue-green precipitate	Clear after 5 minutes, clear when heated	No reaction
Secondary alcohol	Blue-green precipitate	Cloudy reaction in 3-5 minutes or when heated	No reaction
Tertiary alcohol	No reaction	Cloudy within a minute	No reaction
Aldehyde	Blue-green precipitate	No reaction	Yellow-orange
Ketone	No reaction	No reaction	Yellow-orange

$$R-\overset{\overset{\displaystyle O}{\|}}{C}-CH_3 \quad \xrightarrow[\text{base}]{X_2} \quad R-\overset{\overset{\displaystyle O}{\|}}{C}-O^- \quad + \quad CHX_3$$

R = H, alkyl, aryl
X = Cl, Br, I

1. When the halogen is iodine, this is the *iodoform reaction*. Note what the reagents are, then predict the result of adding iodine and base to the following substances. Draw their structures and the product of the reaction, if any.

    a. Propanal

    b. Propenone

    c. Butanone

$$RCH_2OH + \text{Jones reagent} \rightarrow \text{blue - green ppt}$$
$$R_2CHOH + \text{Jones reagent} \rightarrow \text{blue - green ppt}$$
$$R_3COH + \text{Jones reagent} \rightarrow \text{no reaction}$$

2. What are the two molecules that react? Why might they react when the third does not? What type of molecule is the nonreactive molecule?

3. What reaction, if any, would you expect from the following molecules, and why?

    a. Ethanol

    b. 2-Methyl-2-propanol

    c. 2-Butanol

MATERIALS

Gloves

Lucas reagent ($ZnCl_2$ in concentrated HCl)

Jones reagent (chromic acid)

2,4-Dinitrophenylhydrazine (2,4-DNP)

Acetone

Ethanol

Test tubes

Tongs

Droppers

Alcohol burner or Bunsen burner, beaker and stand, or water bath

Aldehyde

Ketone

1° alcohol

2° alcohol

3° alcohol

Table sugar

Yeast

250 mL Erlenmeyer flask

PROCEDURE 7.1: IDENTIFICATION OF ALCOHOLS

1. Label four test tubes "1° alcohol," "2° alcohol," "3° alcohol," and "unknown." Add 5 drops of each compound to the test tubes.

2. Add 10 drops of acetone and 2 drops of Jones reagent to the tubes and record the results. A blue-green color or precipitate is a positive reaction.

3. Repeat the experiment, adding 15 drops of Lucas reagent to each test tube. If no reaction occurs, heat the reaction in a water bath using tongs.

POST-LAB QUESTIONS

1. Using the results of the above experiments, what is the identity of your unknown? Explain.

2. What would the results have been if it had been either of the other two molecule types?

PROCEDURE 7.2: IDENTIFICATION OF ALDEHYDES AND KETONES

1. Label four test tubes "1° alcohol," "aldehyde," "ketone," and "unknown." Add 5 drops of each compound to the test tubes.

2. Add 10 drops of acetone and 2 drops of Jones reagent to the tubes and record the results. A blue-green color or precipitate is a positive reaction.

3. Repeat the experiment, adding 10 drops of each compound, 5 drops of ethanol, and 10 drops of 2,4-DNP to each test tube. Record your results.

POST-LAB QUESTIONS

1. Using the results of the above experiments, what is the identity of your unknown? Explain.

2. What would the results have been if the unknown had been either of the other two molecule types?

PROCEDURE 7.3: ENZYMATIC ETHANOL PRODUCTION

1. Place about 5 grams of sucrose in a 250 mL Erlenmeyer flask. Add warm water (~80 °F) to the 100 mL line, and add a generous pinch of yeast.

2. Gently swirl the mixture, and allow it to react for as long as possible.

3. While you are waiting for the yeast and sugar to react, design a protocol using Jones reagent, Lucas reagent, and 2,4-DNP to show that the product of the reaction is a primary alcohol, not an aldehyde or a ketone.

4. Execute the protocol and record your results. Decant the liquid carefully, to get as close to a clear liquid as is possible.

POST-LAB QUESTIONS

1. What evidence do you have that a reaction is occurring?

2. How did you demonstrate that one of the results of the reaction is a primary alcohol?

Chapter 8

Mixtures, Solution Concentrations, and Diffusion

If you have been keeping up with the material thus far, this chapter is a review in many ways: solution concentrations are nothing more than density calculations with different units. Instead of worrying about grams per milliliter or grams per cubic centimeter, you will now be calculating moles per liter for molarity. There is no real difference in the mathematics, and no reason that it should cause you any concern.

Problems relating to concentrations can also be thought of as unit conversions: the conversion from g/mL to mol/L or even to ppm or ppb will use the same techniques you used earlier in the book to convert inches to kilometers, or kilograms to ounces. Proceed with caution, and all will be well.

The helpful phrase "like dissolves like" governs the interactions between the solute and solvent molecules. As your chapter explains, there are numerous ways that solutes and solvents can interact, from polar molecules interacting with one another's dipoles, to ions interacting with solvent dipoles, to nonpolar molecules forming hydrophobic interactions that stabilize the molecules near each other. With what you already know of these interactions, the ability of different substances to mix with one another can be predicted and tested. One of the most interesting effects you will see is liquids that cannot mix with one another arranging themselves in layers by density. Substances that we think of as very soluble will sit unaffected at the bottom of a beaker if they are unable to mix with the liquid above them

Solutions are not always as simple as salt or sugar dissolving in water. What, exactly, for instance, is gelatin? The collagen molecules in gelatin powders or leaves are very hydrophilic (leave some of the powder out on a humid day, and you will get a demonstration of this). When the gelatin molecules are boiled, they straighten out, and water molecules can attach to every polar spot. If the boiling was complete, when the gelatin cools the long collagen molecules keep holding onto the water molecules, making a peculiar substance that is not quite liquid and not quite solid. It can be left as long as you like and it will never settle out, though eventually it will dry to a flat sheet. How can you know this is a colloid? Even though colloids look clear, these large molecules scatter light. If you shine a flashlight at a glass filled with food coloring and sugar water, the light will more or less pass right through. Do the same thing with a glass filled with water, food coloring, and water-bound collagen, and the light will scatter: there is no path through that substance that will allow the light to pass straight through.

So, do colloids have any practical applications? Mayonnaise, paint, toothpaste, firefighting foam, hand lotion are all examples of colloids. Despite the fact that they are not very familiar, colloids are a part of everyday life. Now that you know what you are looking for, you will see them all around you.

1. Each of the following substances will exhibit a primary intermolecular force when added to a mixture or solution. Identify the primary intermolecular force the following substances will exhibit:

    a. NaCl

    b. Water

    c. Sucrose

    d. Hexane

    e. Isopropyl alcohol

    f. Naphthalene (see diagram)

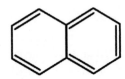

2. Predict the results of the following mixtures. You will test your predictions in the lab.

    a. Water + NaCl

    b. Water + Sucrose

    c. Water + Naphthalene

    d. Water + Hexanes

    e. Water + Isopropyl Alcohol

    f. Isopropyl alcohol + NaCl

    g. Isopropyl alcohol + Sucrose

    h. Isopropyl alcohol + Naphthalene

    i. Isopropyl alcohol + Hexanes

    j. Hexanes + NaCl

    k. Hexanes + Sucrose

    l. Hexanes + Naphthalene

3. Look up or calculate the molecular weights of NaCl, sucrose, and casein.

MATERIALS

Water

Hexanes

Isopropyl alcohol

Sodium chloride

Sucrose

Naphthalene

Small beakers or test tubes

25,000 MW cutoff dialysis tubing

Dialysis clips (optional)

Powdered milk

Food coloring

250 mL beaker

Stirplate

Vinegar

Baking soda

PROCEDURE 8.1: INTERMOLECULAR FORCES

1. Wearing gloves and working cautiously, mix the following substances to see what the results are. Do the solids dissolve? Are the liquids miscible?

    a. Water + NaCl

    b. Water + Sucrose

    c. Water + Naphthalene

    d. Water + Hexanes

    e. Water + Isopropyl alcohol

f. Isopropyl alcohol + NaCl

g. Isopropyl alcohol + Sucrose

h. Isopropyl alcohol + Naphthalene

i. Isopropyl alcohol + Hexanes

j. Hexanes + NaCl

k. Hexanes + Sucrose

l. Hexanes + Naphthalene

2. Record your data in a table.

POST-LAB QUESTIONS

1. Which of the mixtures will conduct electricity? Why?

2. What are the primary intermolecular forces within water, hexanes, and isopropyl alcohol? Support your answer using the formula and structure of each molecule.

3. What types of bonding are seen within NaCl, sucrose, and naphthalene units?

4. In the mixtures in which the substances dissolved (or mixed completely, for liquids) what intermolecular forces were responsible for that dissolution or mixing?

5. For liquids that did not mix, which formed the top layer, and which formed the bottom layer? Which liquid was the densest? The least dense?

PROCEDURE 8.2: DIALYSIS AND OSMOSIS

1. Take four pieces of dialysis tubing of equal lengths and soak them until they are pliable.

2. Close one end, either with a tight knot or a dialysis clip.

3. All of the solutions available have the same concentrations in mg/mL. Using a dropper, fill each dialysis tubing to the same height with a solution of:

    a. Sodium chloride

    b. Sucrose

    c. Powdered milk (casein)

    d. Food coloring

4. Close the other end of the tubing, again using either a tight knot or a dialysis clip, and place the filled tubing into beakers of pure water. Use the largest available beakers, and if possible, place each of the tubes in its own beaker.

5. Observe the changes in the tubes and water over the course of the lab period.

6. Record whether water appears to be moving into the dialysis tubing, out of the dialysis tubing, in both directions, or in neither direction.

POST-LAB QUESTIONS

1. Which piece of tubing showed the most dramatic effect?

2. For each of the experiments, did water flow into or out of the tubing?

3. What happened with the tubing containing food coloring?

4. Given the weights of the known compounds, what is occurring in the tubes: osmosis, dialysis, or both? Why?

5. Proteins are often packed in salt. Could this tubing be used to separate salt and a protein? Why or why not?

PROCEDURE 8.3: MAKE YOUR OWN COLLOID

1. Place about 100 mL of vinegar in a 250 mL beaker, and mix in about 1 mL dishwashing liquid and a drop or two of food coloring, if you like.

2. In a separate beaker, mix about 7.5 g baking soda with 100 mL of water.

3. Mix the contents of the two beakers, and observe the results.

POST-LAB QUESTIONS

1. What is being produced in this reaction?

2. Is the colloid (or, more accurately, foam) produced by this reaction stable? What evidence do you have for this?

3. Firefighters use a variation on this reaction to fight fires. Why would this be more effective than simply spraying a fire with water?

Chapter 9

Acids and Bases

Identifying acids and bases is relatively straightforward. Acids taste sour, while bases taste bitter. Acids turn blue litmus paper red, and bases turn red litmus paper blue. When acids and bases are mixed, they provide us with water and salt. What is surprisingly not straightforward is the chemical nature of acids and bases. What is it that makes it possible to look at a chemical structure and say, "That's an acid"? This was a problem that confounded scientists nearly as long as the nature of the atom had.

A misconception that lasted a tremendously long time was that acids all contained oxygen. Consider the definitive acid that everyone knows, HCl, which is used abundantly in this lab. No oxygen there. So, the definition had to be re-worked. The most common working definition at present of an acid is that it is a compound that will donate a hydrogen into solution. This makes it a little easier to look at a formula and see what it's going to do: HCl in water dissociates to form hydrogen ions and chloride ions, so it's an acid.

On the other hand, ammonia, NH_3, is considered a base. It removes hydrogens from solution to form ammonium ions, NH_4^+. While it may seem contradictory that ammonia is considered a base even though it lacks hydroxides, it is possible because ammonia removes hydrogen from solution to form ammonium ions, making it a base.

For the introductory chemistry student, it is often advisable to "memorize this list of acids and bases" until you have a firm handle on which substances will donate a hydrogen and which substances will cause the removal of a hydrogen.

Perhaps the most mysterious substances in the world of acids and bases are *buffers*, compounds that seem practically sentient in their ability to adapt to changing conditions. In reality, these systems preserve a not-terribly-complex equilibrium according to Le Châtelier's principle: add acid, and the systems become more basic; add base, and they become more acidic. Buffers do *not* maintain neutrality. The system of a weak acid and its salt, or weak base and its salt, will have a specific pH. Once it is out of that range, the buffer will no longer hold the pH steady. It is for this reason that there are so many buffer systems: some will hold a pH as basic as 10, others a pH as acidic as 2.5.

An acetic acid buffer holds a somewhat acidic pH. If acid is added, more of the non-dissociated form appears. If base is added, the hydrogen ions disappear from the solution. For example, to add buffer capacity, a salt is added (in this case, sodium acetate) so that there is more of the (acetate) ion to be pushed and pulled around, swinging from being acetic acid to acetate ion as needed to hold the pH steady.

$$CH_3COOH \rightleftharpoons CH_3COO^- + H^+$$

1. What is the molarity of sodium hydroxide in a solution prepared by dissolving 50 g of NaOH in sufficient water to make 1500 mL of solution?

2. What is the pH of a solution with an H^+ ion concentration of 10^{-5} M? An OH^- concentration of 10^{-4} M?

3. How much 1 M NaOH is needed to neutralize 10 mL of 6.5 M HCl?

4. If 27.5 mL of a 0.1 M NaOH solution is needed to neutralize 50.0 mL of an HCl solution, what was the initial molarity of the HCl solution?

5. What is a buffer? Of what is a buffer made? In the bicarbonate buffer system found in the blood, what is the purpose of the bicarbonate?

6. What does it mean for an acid to be weak? Strong?

MATERIALS

pH 4, 7, and 10 standards

Red cabbage indicator

Vinegar

Ammonia

Bleach

Hydrogen peroxide

Apple juice

Sodium bicarbonate (baking soda) solution

Test tubes

Muriatic (hydrochloric) acid

1 M sodium hydroxide

Stir plates

Phenolphthalein solution

5 mL volumetric pipets (optional)

10 mL graduated pipets

1.0 M acetic acid

1.0 M sodium acetate

pH meter (if available) or pH paper

PROCEDURE 9.1: PH INDICATORS

1. Label test tubes with the pH of each of the available standards. Fill the tube up to the midpoint, and add a few drops of red cabbage indicator. These are now your pH meter: acid, neutral, and base.

2. In the same way, test the pH of the ammonia, bleach, vinegar, apple juice, hydrogen peroxide, and anything else your instructor provided.

POST-LAB QUESTIONS

1. Rank the substances from most acidic to most basic.

PROCEDURE 9.2: TITRATION

1. *Muriatic acid* is the common name for hydrochloric acid (HCl). It can be bought at the hardware store. Put a stirbar in a 250 mL beaker. Carefully pipette 5 mL of muriatic acid into 50 mL of water. It is important to put the acid into the water, as the mixing of the water and acid releases heat.

2. Add a drop or two of phenolphthalein solution to the acid solution.

3. Turn on the stirplate, and begin pipetting in 1 M NaOH with a graduated pipet. Keep careful track of how many milliliters you have added: it will require multiple pipets, so don't worry if your count is getting high.

4. When the color change in the solution starts to linger, slow down your additions. When the color change is permanent, your titration is complete.

POST-LAB QUESTIONS

1. Using the initial volume of acid (5 mL) and the amount of 1 M NaOH required to reach the endpoint of the solution, calculate the molarity of the muriatic acid solution.

2. What does the phenolphthalein do? Why is it added?

3. The water added at the beginning of the titration is meant to provide a measure of safety. Why does its volume not need to be considered a factor in the titration?

PROCEDURE 9.3: BUFFERS

1. Make up a buffer solution by mixing with a stirbar 100 mL of 1 M sodium acetate and 100 mL acetic acid in a 250 mL beaker.

2. Measure the pH of the solution using a pH meter, pH paper, or add a drop of the cabbage indicator. Record the pH.

3. Put 200 mL of water and a stirbar in another beaker and measure the pH using the same method as above.

4. Put the water on a stirplate, and add 10 drops of 1 M hydrochloric acid, measuring the pH between each drop or recording the color change after each drop.

5. Add 10 drops of 1 M NaOH, recording the pH between each drop. Add 10 more drops, again recording the pH between each drop.

6. Repeat the entire procedure with the buffer solution.

POST-LAB QUESTIONS

1. Using the concept of equilibrium, explain the results of adding acid and base to a buffer.

2. Is it possible to add enough acid or base to cause a dramatic change in the pH of this solution? Why or why not?

Chapter 10

The Reactions of Organic Functional Groups in Biochemistry

Electrons move around a lot in reactions. In inorganic chemistry, the reasons electrons behave as they do are beyond the scope of this class. In biochemistry, electrons move around a lot, too. Electrons being passed down an energy gradient is ultimately what drives the production of ATP, something that will become very important in later chapters. This is important to you personally right now: any disruption in the workings of that electron transport chain, whether stopping it or removing the controls that keep it running smoothly, will kill you quite promptly.

We make reference to "burning" calories, but how common is it for us to really think about what this means? Through hydrolysis reactions, oxidation-reduction reactions, and all the other reactions that are collectively termed "catabolism," our bodies combine oxygen with the molecules of the food we eat and "burn" the food into carbon dioxide, water, and energy, just as an automobile does. The radical difference between the way we consume energy and the way an automobile does is the level of control that our digestive enzymes exert over the process. When our bodies burn fuel, as they do very slowly (with lots of the reactions seen in this chapter), the heat released is used to maintain our body temperatures under very tight control. When a car burns fuel, the inside of the engine becomes unbearably hot very quickly. There are poisons, such as rotenone and cyanide, that will uncouple the chain of reactions by which food is "burned" to create metabolic energy. It should come as no surprise that one of the symptoms of these poisons is an extremely high fever.

As our bodies work, various systems are broken down, recycled, and built back up in reactions of various organic functional groups. Polymers are built up and broken down, and damaged molecules are taken apart and replaced. What is remarkable is the accuracy with which this occurs, and the precision with which these molecules can be identified. In lab, you will watch electrons moving around, identify organic functional groups, and perform a biochemical reaction referred to as saponification with chemical methods that, compared to the cascades of enzymes that so elegantly accomplish tasks like these in the body, can only be described as brute force.

Also remarkable is how important water is in these biochemical reactions. Far from simply being the carrier in which these molecules move, water is an integral part of what is going on. Hydrolysis, dehydration, hydration … all of these reactions have water not just as a carrier but as an integral reactant.

PRE-LAB EXERCISE

1. Metals will react with each other to pass electrons back and forth. In the following reaction, identify the oxidizing agent and the reducing agent.

$$2\ Na + Cl_2 \rightarrow 2\ NaCl$$

2. Look up the structures of phenol, salicylic acid, acetylsalicylic acid, acetaminophen, and ibuprofen. Draw them here, and identify each of them as being an aromatic alcohol (phenol) or not.

    a. Phenol

    b. Salicylic acid

    c. Acetylsalicylic acid

    d. Acetaminophen

    e. Ibuprofen

MATERIALS

Aluminum foil

Copper (II) chloride (cupric chloride)

Glass rod

250 mL beaker

Copper wire

0.1 M silver nitrate solution

Test tubes

Salicylic acid

Aspirin

Acetaminophen

Ibuprofen

Mortar and pestle

2.5% ferric chloride solution

4% phenol solution (optional)

Lard

Sodium hydroxide

Sodium chloride

Methanol

Stir rods

PROCEDURE 10.1: OXIDATION AND REDUCTION

COPPER AND ALUMINUM

1. Take a 100 mL beaker and add a small scoop of copper chloride crystals (but not enough to completely cover the bottom). Add water up to about the 100 mL mark, and stir with a glass rod until completely dissolved.

2. Take a small piece of aluminum foil, about 4 inches square, crumple it lightly, and hold it under the copper chloride solution with the glass rod.

3. Observe the reaction until it is complete, that is, there is no more bubbling and either the aluminum or the color in the test tube is gone.

POST-LAB QUESTIONS

1. Was there a difference in the color of the copper chloride and the copper chloride solution? What might account for this, if there was a change?

2. What was happening with the aluminum foil?

3. What was happening with the copper chloride solution?

4. Write and balance the reaction showing only the metals, and make sure the charges are balanced.

SILVER AND COPPER

1. Take a 6-inch piece of copper wire and wind it into a loose spiral around a pencil. Drop it into a test tube, and cover it with silver nitrate solution.

2. Observe the reaction as long as possible.

POST-LAB QUESTIONS

1. What was the final color of the solution around the wire? What might be causing this color?

2. What was happening to the copper wire? What was deposited on it?

3. Write and balance the reaction showing only the metals, and make sure the charges are balanced.

PROCEDURE 10.2: IDENTIFICATION OF PHENOLS

1. Grind up a small sample of salicylic acid, a tablet of aspirin, a tablet of acetaminophen, and a tablet of ibuprofen. Carefully clean the mortar and pestle between samples to make sure that there is no cross-contamination.

2. Put a sample of each substance in a labeled test tube, and add water to the mid-point of the tube.

3. Add the same amount of water to a fifth tube, and label it "negative control." Take a sixth tube, add water and a few drops of phenol solution to it, and label it "positive control."

4. Add 10 drops of ferric chloride solution to each test tube, and note the color of each tube.

POST-LAB QUESTIONS

1. Using your answers from the pre-lab, did the ferric chloride react with all of the phenols?

2. Were the reactions identical, or were there differences in them? If there were differences, what were they?

3. Was there a reaction with aspirin? If so, what would be causing it?

PROCEDURE 10.3: SAPONIFICATION

1. Place about 12 g of lard in a 150 mL beaker.

2. Carefully make a solution of 5 g sodium hydroxide in 15 mL water.

3. Add the sodium hydroxide solution and 10 mL of methanol to the lard. Stir well.

4. Prepare a solution of 50% methanol and water, and bring some in a 50 mL beaker to your work area (you may need to get more when the solution runs out).

5. Put the lard, sodium hydroxide, and methanol on a hot plate or over a very small flame, and heat with constant stirring so that it boils but does not splatter. Do not allow it to boil dry. As needed, add some of the water/methanol mixture.

6. In a 500 mL beaker, prepare a strong salt solution containing 100 g of sodium chloride in 300 mL water.

7. When the saponification reaction is complete, there will no longer be any oily globules, and the mixture will no longer smell like fat.

8. Place a piece of filter paper in a funnel and set aside.

9. Pour the hot saponification mixture carefully into the salt solution and stir vigorously.

10. Collect the precipitated soap in the filter, wash with cold water, then scrape it into a paper cup (or two) and allow it to dry before using.

POST-LAB QUESTIONS

1. Was there a distinct difference in the properties of the substances you used before and after you performed this reaction?

2. Would you expect the process to proceed the same way with different fats? Why or why not?

3. Soap allows otherwise insoluble oils to become soluble in water. Using the principle "like dissolves like" and the basic structure of a soap molecule below, explain how soaps allow the mixing of polar and nonpolar substances.

Chapter 11

Carbohydrates: Structure and Function

Carbohydrates divide fairly neatly into three groups: monosaccharides, disaccharides, and polysaccharides. There are numerous ways to identify which saccharide you have, and a number of ways to convert from one to another. If you enjoy solving a puzzle, this is just a small sample of what the study of chemistry and biochemistry has in store should you choose to continue to study it. With a little logic and an understanding of the structures you are dealing with, you can learn a great deal of information about things you can't see.

Because there is so much information to be gained qualitatively, there's a lot of "about" and "approximately" in biochemistry. As an example, iodine provides a distinctive color change in the presence of starch. The amount of iodine and the amount of starch are not all that important. If you see the color change, you know that there is starch present. But if you start by weighing out a specific quantity of starch and making a solution of known concentration (w/w), it is possible to dilute this solution over and over again until the distinctive color of iodine disappears. By calculating the concentration of the last sample in which the color was visible, it is possible to determine how sensitive this method is, and how much starch is needed in solution before it can be detected this way.

What is also possible to observe is the specificity with which enzymes function. It is remarkable that we can digest starch but not cellulose. The differences between the two require a careful explanation to see how they are chemically different. How enzymes do what they do is a subject for a later chapter: here we are simply using them as detection tools. An enzyme's ability to distinguish one polymer from another, and even one disaccharide from another, makes them remarkable tools for identifying the contents of a solution.

Carbohydrates themselves are pretty remarkable molecules, beginning with the fact that they all come pretty close to sharing the empirical formula that gives them their name: CH_2O. Like other topics in chemistry, there were some misconceptions along the way, but the description of carbohydrate molecules as "hydrated carbon" has its advantages, at least for the student having to identify them. Although they are not our most efficient source of energy—an honor that goes to fats, at 9 calories per gram—they are our fastest source of energy. The labs in this chapter will demonstrate how well equipped the human body is to use carbohydrates in all forms. We have enzymes that begin digesting polysaccharides the moment we ingest them, and mono- and disaccharides are able to pass immediately through the mucous membranes that line our digestive tracts, making it possible for a diabetic to feel the benefit from a glucose tablet immediately when letting it dissolve in the mouth.

Fats can make no such claims: not being water soluble means that extensive digestion is needed for them to provide any energy benefit. No one ever put a pat of butter under their tongue for some quick energy, or if they did, it probably didn't work!

PRE-LAB EXERCISE

1. Draw the open-chain structure of glucose, the structure of sucrose, the structure of lactose and a unit of starch. Identify any aldehyde groups.

    a. Glucose

    b. Sucrose

    c. Lactose

    d. Starch

MATERIALS

Iodine solution

Benedict's solution

Soluble starch

Glucose

Sucrose

Gelatin

Test tubes

Hot water bath

Concentrated HCl

NaOH solution

Lactose

Lactase tablets

Invertase or sucrase

Glucose test strips

PROCEDURE 11.1: TESTING FOR STARCH AND SUGAR

1. Using 10 test tubes, create two of each of the following five labels: starch, monosaccharide, disaccharide, protein, and water. These will be used with two different test solutions.

2. Make 150 mL of 2% starch and 2% gelatin solutions. Also, make 50 mL solutions of 2% glucose and 2% sucrose. Record how these solutions were made.

3. To assay for starch, fill a test tube halfway with each solution, place 1 drop of iodine solution in each tube, and record the results.

4. To assay for aldoses, fill each test tube about one-third full, add 5 drops of Benedict's solution, and place in the hot-water bath for 10 minutes. Record the results.

5. To determine the sensitivity of the starch test, take 10 mL of the original solution, and begin diluting by adding 1 mL of starch to 9 mL of water, then adding a drop of iodine. Continue 10x dilutions until the blue color can no longer be seen.

POST-LAB QUESTIONS

1. Make a table with the molecule tested, the molecule type, and the result of the reaction.

2. Starch and proteins are both long-chain polymers. Are there tests to distinguish between these two polymers? Is the iodine test definitive for polymers or for starch?

3. How many times did you need to dilute the starch solution before you could no longer detect the blue of the starch? If the original solution was 1 part of starch per 50 parts of solution, what is the concentration at which the starch cannot be detected by this test?

PROCEDURE 11.2: CONVERTING STARCH TO SUGAR

1. Take 50 mL of the starch solution made previously, and add 5 drops of concentrated hydrochloric acid. Put this in a hot water bath for 30 minutes, cool, and neutralize with NaOH solution.

2. Repeat the above with 50 mL of the gelatin solution made previously.

3. At the same time, place 50 mL of the starch solution in an Erlenmeyer flask and add 1–2 mL of fresh saliva (you read correctly). Put in a warm water bath for 20 minutes or longer.

4. Repeat the above with 50 mL of the gelatin solution made previously.

5. To assay for starch, fill a test tube halfway with each solution, place 1 drop of iodine solution in each tube, and record the results.

6. To assay for aldoses, fill each test tube about one-third full, add 5 drops of Benedict's solution, and place in the hot water bath for 10 minutes. Record the results.

POST-LAB QUESTIONS

1. Make a table recording the molecule tested, the molecule type, and the result.

2. Which was more effective, acid hydrolysis or enzyme hydrolysis? Justify your answer with the results of the tests above.

3. What information could you obtain from the tests on gelatin? Does this demonstrate that the acid hydrolysis was unsuccessful?

PROCEDURE 11.3: ENZYME SPECIFICITY

1. Label four test tubes "lactose." Add approximately 0.3 g lactose to each, and add 10 mL water.

2. Label four test tubes "sucrose." Add approximately 0.3 g sucrose to three other test tubes, and add 10 mL water.

3. Add 10 mL water to four test tubes as controls.

4. Add powdered lactase to one of each of the lactose, sucrose, and water tubes.

5. Add powdered invertase to one of each of the lactose, sucrose, and water tubes.

6. Add 5 drops of concentrated HCl to one of each of the lactose, sucrose, and water tubes.

7. Allow the enzyme tubes to sit for about 10 minutes, and test each with glucose test strips.

8. Put the acid-containing tubes in a water bath for 30 minutes, then cool and neutralize with NaOH before testing with glucose test strips.

9. If there is time, repeat the procedure with milk, soda, or other products to determine what sugars are present.

POST-LAB QUESTIONS

1. Make a table recording the molecule tested, the molecule type, and the result.

2. What do the results of this experiment tell you about the specificity of enzymes compared to chemical hydrolysis?

3. The stomach contains hydrochloric acid, but it is possible to be lactose intolerant or, more rarely, to be sucrose intolerant. What does this tell you about the digestion of sugars in the human body? Are they digested with acid only, enzymes only, or something in between? Support your answer.

Chapter 12

Lipids: Structure and Function

Lipids are unique among the classes of molecules we will study. While carbohydrates are classified by structure, proteins are organized by the way their specific units are attached together, and nucleic acids are identifiable by their few monomers working together through types of bonding not seen anywhere else, lipids are grouped together on the basis of something quite straightforward—they are insoluble in water and soluble in organic solvents. That is not all that much to go on compared to other biomolecules that are characterized by structure and function.

But, it turns out that this is enough. At this point, you should be able to recognize a hydrophobic molecule when you see one, and that is the same thing as a molecule that's soluble in polar solvents. Lipids are very diverse, but they are also the only molecules that are discussed, structure and all, as a matter of routine in health news.

The bad guy for many years was cholesterol, a lipid found in cell membranes that the body produces in response to inflammation. Cholesterol is a protective molecule that is essential for life. It is synthesized primarily in the liver and the brain, and it is poorly absorbed from food, but for years, people were taught not to eat eggs because of their high cholesterol levels. The molecule has a role in cell signaling, is a primary component of the myelin sheath that insulates nerve cells, and is a precursor to bile, vitamin D, cortisol, aldosterone and all of the sex hormones. Newly released medical guidelines suggest that perhaps the focus on cholesterol in the war against cardiovascular disease may be misplaced, and that the emphasis on lowering cholesterol, which may act as an antioxidant, may not be as important as was originally thought.

So, what is a scientifically educated consumer to think? The vast complexity of the human body, in fact of any living species, makes medical science just as subject to fads as any other field of endeavor. Parents can tell you that medical opinion was quite clear on the care of infants—they should always be put to sleep on their stomachs—until they abruptly changed course, and all infants were to be put to sleep on their backs.

Do we throw up our hands and conclude that scientists and doctors have no idea what they are talking about? No, in this area, as in every other, there is no substitute for common sense. If a food doesn't occur naturally, for example, *trans* fats, it is unlikely to be good for us. In this case, science figured out a health puzzle: margarine was touted as a health food, but eating it increased the risk of heart attack. If a food occurs in nature and has been eaten by humans for thousands of years, like milk or eggs, chances are it is okay. How about an all-bacon diet? That's probably a bad idea. If it's provided only moderately by nature, like bacon and alcohol, it should probably only be consumed moderately. When in a state of overall good health, as most students are, we all have a good sense of what makes us feel better, like a healthy diet, and what makes us feel worse. Our bodies have ways of telling us how we are doing overall that don't require blood tests, just that we pay attention to how we feel and think twice about fads in every field, from food science to medicine.

1. Look up the primary components of the following natural fats: butter, olive oil, lard, coconut oil, and vegetable oil. Identify them as saturated, monounsaturated, or polyunsaturated. From their structures and degree of saturation, predict their relative melting points.

    a. Butter
        (i) Primary component

        (ii) Degree of saturation

        (iii) Melting point ranking

    b. Olive oil
        (i) Primary component

        (ii) Degree of saturation

        (iii) Melting point ranking

    c. Lard
        (i) Primary component

        (ii) Degree of saturation

        (iii) Melting point ranking

    d. Coconut oil
        (i) Primary component

        (ii) Degree of saturation

        (iii) Melting point ranking

    e. Vegetable oil
        (i) Primary component

        (ii) Degree of saturation

        (iii) Melting point ranking

MATERIALS

Butter

Olive oil

Lard

Coconut oil

Vegetable oil

Soap

Detergent

Stearic acid

Cetyl alcohol

Lanolin

Triethanolamine

Glycerin

Commercial hand lotion

100 mL beakers

10 mL graduated cylinder

50 mL graduated cylinder

pH paper

PROCEDURE 12.1: PHYSICAL PROPERTIES OF LIPIDS

1. Take samples of butter, vegetable oil, olive oil, coconut oil, and lard out of the freezer. All of the samples should be solid. Using your predictions from the pre-lab, place the two samples with the highest melting point in a water bath and leave the other two on the lab bench.

2. Observe the samples as they melt, and use a thermometer to determine the approximate melting point of each substance.

POST-LAB QUESTIONS

1. Rank the substances tested in order of melting point. Did your predictions match the melting points you observed? Account for the matching or for any differences you observed.

PROCEDURE 12.2: PROPERTIES OF DETERGENTS

1. Make a solution of 1 gram of soap and 100 mL of water, and another solution of 0.5 grams of detergent and 100 mL of water.

2. Put 5 mL of each solution (soap or detergent) into five test tubes, for a total of 10 test tubes. Into an 11th test tube, place 5 mL of distilled water as a control. To each test tube, add 3 or 4 drops of each melted fat or oil and shake vigorously, either stoppering the end of the test tube or covering it carefully with a gloved thumb.

3. Describe the results. Do you see emulsification? Do the fat and liquid mix, or do they return to being separated? Record your results in a table.

4. Using a solution of 2% bromine, test for unsaturation of the soap and detergent solutions. What does this tell you about the action of these compounds? Are they breaking chemical bonds or simply permitting mixing? Justify your answers.

POST-LAB QUESTIONS

1. What do the results of the unsaturation tests tell you about the action of the soap and detergents? Are they breaking chemical bonds or simply permitting mixing? Justify your answers.

PROCEDURE 12.3: INGREDIENTS IN HAND LOTION

1. Place the following nonpolar ingredients in one beaker:
   a. Stearic acid   3 g
   b. Cetyl alcohol   1 g
   c. Lanolin   2 g

2. In another beaker place the following polar ingredients:
   a. Glycerin   2 mL
   b. Water   50 mL
   c. Triethanolamine   1 mL

3. Place both beakers in a water bath until all ingredients are melted and warm. Carefully combine the contents of the beakers, pouring the water and glycerin into the other ingredients, and stirring with a glass rod until a smooth lotion is obtained. Using a piece of pH paper, test the pH of the lotion.

4. Repeat the procedure, omitting the triethanolamine. What differences do you see in the finished product? What is the pH of the final product?

5. Repeat the procedure, omitting the stearic acid. What differences do you see in the finished product? What is the pH of the finished product?

POST-LAB QUESTIONS

1. Based on what you observed, what is the purpose of each ingredient in the following list?
   a. Triethanolamine

b. Stearic acid

2. Look up the structures of these molecules, and write an acid/base equation for their reaction.

Chapter 13

Proteins: Structure and Function

Proteins are the workhorses of the cell. They have structural, mechanical, catalytic, signaling, adhesive, immunological, storage, transport, cell cycle and nutritive functions. Of all the macromolecules, proteins have the largest number of monomers—20 amino acids occur in nature—and the greatest diversity of structures. From the smallest peptide hormones of three amino acids, to large, multi-subunit glycoproteins, the importance of proteins was recognized well before the importance of nucleic acids, and nucleic acids are not able to function without the assistance of proteins.

The primary structure of proteins is fairly simple: amino acids linked together by peptide bonds. But, the differences in the 20 amino acids are pronounced enough that they are fairly simple to separate, something that will be done during these exercises. Even if proteins consisted only of straight chains of amino acids, an almost infinite number of proteins would be possible, and each animal and plant has species-specific versions even of proteins that are ubiquitous in cells. Once the amino acids join into peptides and then proteins, the process of separating them from one another becomes an increasingly difficult problem. To locate and purify a single protein from the inside of a cell without destroying its structure requires a sophisticated and delicate effort.

Part of the process of folding a protein is collecting amino acid side chains on the outside of the protein that are suited to the environment in which the protein will function. A protein that will be in the cytoplasm will have soluble, polar proteins on the exterior. Proteins that will function in the center of the membrane will have nonpolar amino acids on the outside, and proteins that will operate as shuttles on the surface of the membrane will have patches of nonpolar amino acids on the surface. Proteins with nonpolar surfaces or with patches of non-polarity are particularly challenging to purify, as the nonpolar surfaces will try to exclude water and stick together. Trying to get the proteins into an aqueous solution without aggregating frequently causes their three-dimensional structure to come apart.

The three-dimensional structure of a protein is essential to its function. A collagen fiber that is broken in pieces or an enzyme with a slightly denatured structure that reduces the catalytic function of the active site are no longer useful. Proteins in living systems do not have an indefinite shelf life: they are subject to oxidative damage, enzymes can be irreversibly inhibited, and cells routinely produce more proteins to replace those that have worn out. But, what does a cell do with these worn-out proteins? This question was partly answered by the winners of the 2004 Nobel Prize in chemistry. When a cell detects a no-longer-useful protein, it is "tagged" with ubiquitin, another protein! This tag signals another cell structure called a proteasome. The tagged and ubiquitinated protein is then chopped into little pieces that can then be reused.

The variety and complexity of the proteins that make up even something as simple as a bacterium is astonishing. The variety of functions that proteins must serve, from protecting us and providing us structure to moving us from place to place, is something that a course of this kind can barely touch on but has provided a lifetime of study for many a scientist.

1. Look up the molecular weights and polarities of the 20 common amino acids and rank them in two lists: one in order of size and the other in order of polarity.

   a.

   b.

   c.

   d.

   e.

   f.

   g.

   h.

   i.

   j.

   k.

   l.

   m.

   n.

   o.

   p.

   q.

   r.

   s.

   t.

2. What effect does exposure to extreme conditions of heat or pH have on proteins?

3. Ninhydrin detects primary and secondary amines. Using structures, explain why it is useful in the detection of amino acids.

4. In the biuret test, copper sulfate binds to peptide bonds. Given this information, could it be used to determine the percent of protein of a solution, the molarity of protein solution, or both? Explain your answer.

MATERIALS

Chromatography paper

Glass capillaries

1000 mL beaker

Glass rod

Tape

Food coloring

70% isopropyl alcohol

50 mL graduated cylinder

Long-stemmed funnel

0.1 M amino acid solutions (three different)

Unknown mixtures of amino acids (0.1 M of two)

2% ninhydrin in ethanol

Milk

Eggs

Hotplate

White vinegar

Gelatin

Test tubes

Test tube holder

1% NaOH solution

5% $CuSO_4$ solution

PROCEDURE 13.1: SEPARATION AND IDENTIFICATION OF AMINO ACIDS

1. Cut a piece of chromatography paper that will fit comfortably into a 1000 mL beaker: approximately 10 cm wide x 15 cm tall.

2. Using a pencil and a ruler, draw a line approximately 2 cm from the end of the paper, and mark 4 spots on the line. Label the spots to indicate the food colors you are going to use: B, R, Y, and G, for example.

3. Use a glass capillary to pick up some of the food coloring. Using a spare piece of chromatography paper, practice making small dots. When you can consistently make a dot of about 2 mm in width, spot the chromatography paper with each of the colors. (Too much will cause the colors to bleed into one another.)

4. Tape the chromatography paper to a glass rod for suspension into the 1000 mL beaker. Adjust the length of the paper so that it does not touch the bottom of the beaker but is suspended slightly above it.

5. Using a long-stemmed funnel, carefully place about 50 mL of 70% isopropyl alcohol in the bottom of the beaker. The solvent should touch the bottom of the paper but not reach up to the spots of food coloring.

6. To prevent excessive evaporation of the solvent, cover the top of the beaker with aluminum foil.

7. Observe the solvent briefly as it travels up the paper. As soon as you are able to see what the moving solvent is doing to the spots of food coloring, continue on.

8. Repeat the procedure from the beginning, this time labeling the spots with the abbreviations for the amino acids you will be using, and "unknown." Spot the paper and suspend it as you did for the food coloring, and again carefully pour in

the solvent. This time, you will not be able to see the progress, so you will need to keep an eye on the progress of the solvent front.

9. When the solvent front is about 2/3 of the way up the paper, carefully remove the paper from the solvent and lay it flat to dry (use a hair dryer to speed the process). When the paper with the amino acids is dry, carefully spray it with 2% ninhydrin, and dry again.

POST-LAB QUESTIONS

1. How many components were in each of the food coloring samples? Were there components that they had in common?

    a.

    b.

    c.

    d.

2. What was in your unknown amino acid sample or mixture?

3. Why should a pencil and not a pen be used to draw the baseline on the paper?

4. Given the distance each amino acid traveled relative to the baseline, develop a hypothesis as to what property it is that allows their separation. Is it polarity? Size? Or something else?

PROCEDURE 13.2: COOKING PROTEINS

1. Using six beakers, put about 50 mL of milk in three 250 mL beakers, and crack an egg into three others.

2. Label one set of milk and egg beakers "heat," another "strong acid," and the last beaker "weak acid."

3. Add 50 mL of water to the two "heat" beakers, and put them on a hotplate until they are well-cooked.

4. Take the "strong acid" beakers and *carefully* add 50 mL of 1 M hydrochloric acid (by way of comparison, stomach acid is roughly 0.15 M but is also almost 100 °F).

5. Place about 50 mL of white vinegar in the beakers labeled "weak acid."

POST-LAB QUESTIONS

1. Of the three treatments, which "cooked" the proteins the most quickly?

2. Based on your observations, what is the process observed in these instances for which we are using the general term "cooking"?

3. Ceviche is a traditional dish in the southern coastal Americas made with raw fish marinated in lemon and/or lime juice (an acid solution roughly equivalent in strength to the vinegar above). Based on this experiment and your observations, is ceviche truly a raw dish? Why or why not?

PROCEDURE 13.3: THE BIURET TEST

1. Carefully make 5 mL of 5%, 2.5%, 1.25%, and 0.625% gelatin solutions by weight, heating the test tubes in a boiling water bath until the gelatin is fully dissolved.

2. Place 5 mL of 10% NaOH, then 250 μL of 5% $CuSO_4$ in each test tube, and mix thoroughly.

3. Make note of the color of the solutions.

4. Make 5 mL of a 20% solution of milk, and repeat the same test.

5. Compare the color of the milk solution to the gelatin solutions.

POST-LAB QUESTIONS

1. Based on the colors of the solutions, what is the approximate percent of protein concentration of the milk solution?

Chapter 14

Nucleic Acid Structure and Function

The number of copies of each chromosome varies widely by species. Humans are diploid, that is, we have two copies of each chromosome. Some frogs are up to dodecaploid, with 12 copies of each chromosome. Other animals, most notably drone bees, are haploid. They have only one set of chromosomes.

Prokaryotes are uniformly monoploid. They aren't referred to as haploid because it is completely normal for them to have only one set of chromosomes. This lack of "backup" in their genomes makes them especially susceptible to mutation; it is usually of a non-beneficial nature, but sometimes it can be a very beneficial one for the organism, which is what we are seeing in the rapid rise of antibiotic-resistant bacteria in hospitals.

If you want to look at DNA, your best bet is to use a species that is easy to manipulate, causes few or no ethical issues, and has lots of DNA. Cultivated strawberries can be up to decaploid, with 10 sets of chromosomes, and most people won't object to squashing a strawberry in order to extract masses of DNA that can be seen by the naked eye.

DNA is a surprisingly hardy molecule: no protein could survive intact the treatment that nucleic acids endure routinely in the process of extraction because nucleic acids have much less complex structures than proteins. If a protein is unfolded, as in Procedure 13.2 there's no simple way to undo that change—you can't un-cook an egg.

In DNA, the covalent bonds holding the strands together are as strong as the covalent bonds holding together the backbone of a protein. The only part that is prone to disruption, as in proteins, are the hydrogen bonds between the strands.

Polymerase chain reaction, or PCR, takes advantage of the relative weakness of the bonds between the strands, and uses unique DNA polymerase enzymes found in bacteria that live in hot springs where the temperatures can reach up to 100 °C. It is PCR that makes "DNA fingerprinting" possible. DNA fingerprinting takes advantage of the great variability in repeated sequences that occur between known genes. The FBI has identified 13 core loci to be used in the United States for genetic identification. A DNA sample from a suspect or a crime scene is "melted" at 100 °C, and little pieces of DNA complementary to the loci being examined are added, along with a DNA polymerase enzyme and a mixture of nucleotides. This mixture is "melted" to separate the strands over and over again, making copy after copy of the areas of interest. The amplified pieces are then separated using a technique very similar to the paper chromatography used in Procedure 12.1.

By using so many areas of the genome that all have such variability within them, the chances of two people sharing the same genetic "fingerprint" are about 1 in 575 *trillion*. As a result, DNA evidence is more likely to exclude the innocent than convict the guilty.

1. In the following exercise, DNA is being extracted in bulk from only one cell type of a strawberry plant. If the entire plant were used, would different sequences of DNA be found in different plant tissues?

2. In DNA fingerprinting, there is the possibility that other DNA contaminants may be found in the mixture, causing erroneous results. If you were purifying the strawberry DNA for genetic analysis, what could you do to ensure that there are no impurities in the sample?

3. Nucleotides can be used not only to carry genetic information but also to carry energy, with their linked phosphate groups. How much energy is provided by the hydrolysis of 1 mol of ATP?

4. If the luciferin/luciferase reaction requires one ATP per photon, and the reaction is 90% efficient, use the formula $E = h \cdot c/\lambda$ to determine the approximate wavelength and color of a firefly flash. $h = 6.626 \times 10^{-34}$ $m^2/k \cdot s$ and Joule $= kg \cdot m^2/s^2$

MATERIALS

Salt

Colorless dish detergent

Water

90% isopropanol (on ice)

Resealable sandwich bags

Strawberries

Graduated cylinders

Filter paper

Glass funnel

Ring stand

Glass rods

50 mL beaker

ATP substrate reagent (0.62 g magnesium sulfate heptahydrate, 55 mg Na_2ATP [1 mM], luciferin, 82.7 mG ATP salt, 185 mg, 1.2 g HEPES buffer, pH 7.5 up to 100 mL with pure water)

Firefly extract diluted, per instructions

Black construction paper

25 mL beakers

Timers

Paper

Index cards

PROCEDURE 14.1: PRECIPITATION AND ISOLATION OF NUCLEIC ACIDS

1. Make the extraction solution by adding 8 mL dish soap and 6 grams of salt to 90 mL of water.

2. Weigh a strawberry, and record its mass.

3. Put a strawberry and about 10 mL of extraction solution in a sandwich bag and completely crush the strawberry. Continue squeezing the strawberry until it is as close to liquid as possible.

4. Suspend a glass funnel in a ring stand and line it with filter paper. Pour the almost-liquid strawberry into the filter paper and allow it to drain into the beaker below.

5. Very carefully pour in an equal quantity of ice-cold 90% isopropanol. Pour the alcohol down the side of the beaker so that it forms a layer on top of the filtered strawberry extract. Do not mix, stir, or agitate the mixture.

6. When a white film develops between the two layers, carefully scoop it up onto a glass rod.

7. Weigh a piece of filter paper, then rinse the DNA strands with more ice-cold isopropanol and lay them on the filter paper to dry.

8. When the paper and DNA are dry, record the weight of the DNA plus paper and then the weight of the DNA alone. Your ability to distinguish between the two will depend on the quality of your scale or balance.

POST-LAB QUESTIONS

1. Where in the strawberries is the DNA found? What is the function of the detergent in the extraction solution?

2. What is the purpose of the filtration step?

3. Why is the cold alcohol not simply stirred into the extraction solution?

4. What is the mass percent of DNA in strawberries?

PROCEDURE 14.2: LUCIFERIN AND ATP

1. Place 1 mL of the firefly extract in a 25 mL beaker. Wrap black construction paper around the beaker to make a dark viewing chamber. Do this experiment twice: the first time, keep the same 1 mL of extract, and the second time get fresh extract between each measurement.

2. Add 1 mL of the ATP solution, and time how long it takes for the light to dissipate.

3. Add 0.5 mL of ATP solution, and time how long it takes for the light to dissipate.

4. Add 0.25 mL of ATP solution, and time how long it takes for the light to dissipate.

5. Add 0.5 mL vinegar to the solution, and time how long it takes for the light to dissipate.

POST-LAB QUESTIONS

1. Graph the results of the experiment with concentration of ATP as the x-axis, and time of light dissipation as the y-axis. Keep in mind that the total volume of the solution determines the concentration. One line should indicate the results when fresh extract was used each time, and another line should indicate the results with the lower concentrations obtained when fresh extract was not obtained.

2. Looking at the graph, was the extract able to regenerate itself between measurements or was more than ATP required?

3. Do the lines have the same slope?

4. Predict the time of light generated by 2 mL of ATP solution added to 1 mL of extract.

5. Did the vinegar generate any reaction? Why or why not?

PROCEDURE 14.3: MUTATION TYPES

1. The following is a small part of the DNA sequence for human insulin, a peptide hormone.

CCATAGCACGTTACAACGTGAAGGTAA

2. Write out the sequence on a piece of paper, and cut a window in an index card that allows three nucleotides, that is, one codon, to show through.

3. Translate the sequence into amino acids from the beginning, using a codon table.

4. Generate a frameshift mutation by beginning at the second nucleotide rather than the first.

5. Generate a substitution mutation by changing one nucleotide in the middle of the sequence. Write the new sequence, along with the new translation.

6. Generate a deletion mutation by removing two nucleotides in the middle of the sequence. Write the new sequence, along with the new translation.

POST-LAB QUESTIONS

1. Rank the three types of mutation by how damaging they are to the resultant protein structure. Justify your answers.

Chapter 15

Energy and Metabolism

Energy and metabolism are huge subjects: in some ways, they are everything when it comes to sustaining life. From the elephants on the African plains to the peculiar microbes of the deep sea, everything has to find energy somewhere and use it to stay alive. The elegance of the multitude of systems and their amazing diversity makes this both a complex subject and a completely fascinating one. In the human body, and in most animal bodies, there are enzymes not only perfectly suited to their tasks but also to their environments. We think of enzymes as being rather delicate objects, if very highly efficient at what they do, and as proof, it is easy to point to the loss of function and denaturation of enzymes, and the loss of structure of proteins, that are seen when outside of their normal pH.

What a small miracle, then, is pepsin, one of the primary digestive enzymes. It is secreted as pepsinogen by the poetically named chief cells of the stomach. As an inactive form and upon hitting the acid conditions of the stomach, it folds into a protein-cleaving enzyme whose first action is to cut itself into an activated form. Most active at a pH of 2, it can survive but not work up to pH 8. Pause for a moment to remember that pH is a logarithmic scale, so this means that whether the concentration of hydrogen ions is 0.01 M or 0.00000001, pepsin will be okay.

The variety of reactions, and their complexity, that allow us to move through not just our days, but the many phases of our lives, are simply astonishing. Before they are born, human babies do not breathe in oxygen. Rather, the fetus pulls in oxygen from the mother's bloodstream with a specialized form of hemoglobin that will never be used after birth. Before they are able to photosynthesize, plants do not breathe in carbon dioxide. Rather, they use oxygen to "burn" the carbohydrate fuel that is packed around them, whether in the form of the flesh of a pea or in the flesh of an apple.

If not used by the seeds as fuel, those same carbohydrates are prone to other forms of oxidative "burning." When a cut apple is left exposed to the air, it quickly turns brown. Because of the slightly dangerous nature of oxygen, our systems also contain any number of molecules known as "antioxidants" that serve in different ways. The enzyme *superoxide dismutase*, grabs oxygen free radicals (O_2^-) and converts them to the far more benign water and hydrogen peroxide. Ascorbic acid (Vitamin C) has the ability to snatch up hydroxyl radicals and incorporate them into its structure.

The burning of fuel, whether wood in a fireplace or glucose in the citric acid cycle, can be dangerous, both for the energy it produces if uncontrolled as well as for its side effects. Much as smoke can be dangerous, so can the side products of the very much slower and more controlled burning of the fuel in our cells. Thankfully, while enzymes are carefully handing off the products of these reactions to one another, other enzymes are scavenging for the damaging "smoke" that they can produce.

PRE-LAB EXERCISE

1. In this lab, you will be provided 0.1 M HCl and 0.1 M NaOH, and you will need to make solutions at pH 1, 3, 5, 7, 9, 11, and 13. Calculate how to produce 100 mL of each of these solutions. Keep in mind that it may be easier to use a previous dilution to make the next: not all solutions have to be made from the initial acid or base and water.

    a. pH 1

    b. pH 3

    c. pH 5

    d. pH 7

    e. pH 9

    f. pH 11

    g. pH 13

2. When seeds germinate, they are not undergoing photosynthesis, but they are instead "burning" carbohydrates with oxygen. Write a formula for the oxidation of glucose.

3. If 2.7 mL of oxygen are consumed by germination over the course of 20 minutes, what is the rate of consumption in milliliters per hour?

4. If the reaction temperature is 30 °C, what is the rate of consumption of oxygen in mol/hour?

5. When cut fruits are exposed to the air, they quickly turn brown. A very similar process occurs in many metals when they are exposed to air and moisture. In the case of metals, what is this process called?

MATERIALS

Gelatin

150 mL beaker

Glass rods

Test tubes

1 M HCl

1 M NaOH

Glass pipets

100 mL volumetric flasks or graduated cylinders

pH paper

Water bath

Ice

Index cards

Dried peas

Dried peas soaked overnight

Plastic beads roughly the same size as the peas

15% KOH solution

Cotton wool

Nonabsorbent cotton

Test tubes

Drilled stoppers

Glass pipettes that will fit through the stoppers

Apples, cut into equally-sized wedges at the beginning of the lab

100 mL beakers containing:

Water

Citric acid solution 0.1%

Ascorbic acid solution 0.1%

Acetic acid solution 0.1%

Acetic acid solution 1%

Tongs

PROCEDURE 15.1: DIGESTION, PH AND TEMPERATURE

1. Heat 3 g of gelatin in 100 mL water until dissolved, and cool to room temperature.

2. While the gelatin cools, prepare seven solutions at the following pH, using your calculations from the pre-lab.

   pH 1 = 0.1 M HCl (provided)
   pH 3 = 0.001 M HCl
   pH 5 = 0.00001 M HCl
   pH 7 = Water, adjust as needed
   pH 9 = 0.00001 M NaOH
   pH 11= 0.001 M NaOH
   pH 13 = 0.1 M NaOH (provided)

3. Make a digestive enzyme solution by adding 10 g of meat tenderizer to 50 mL water.

4. Make three sets of seven test tubes. Label the sets as follows: pH and ice; pH and RT, for room temperature; pH and hot. Each test tube should have a pH and a temperature. Label the fourth set as controls. No enzyme will be added to these, and they will be kept at room temperature.

5. Add 5 mL of the various pH solutions to each test tube. Do this efficiently by measuring 5 mL into one test tube, and marking the level on all the others.

6. As a control, place 5 mL of pH 1, 7, and 13 in three tubes labeled "control." No enzyme will be added to these, and they will be kept at room temperature.

7. Add 5 mL of gelatin solution to each test tube. Stopper and mix, or invert carefully, closing the end with a gloved thumb.

8. Place the "hot" test tubes in a water bath at 37 °C, the "RT" tubes in beaker on the benchtop, and the "ice" tubes in a beaker or water bath full of ice.

9. As efficiently as possible, add 1 mL of enzyme solution to each test tube except those labeled "control."

10. After 20 minutes, examine all the tubes to see under which conditions the enzyme had the greatest effect.

POST-LAB QUESTIONS

1. What conditions were best for the enzyme?

2. Make a table of the results, and develop a scale to indicate the effectiveness of the enzyme under different conditions. It can be a numeric scale (1 to 4) for example, or a color scale, or anything else you can think of to show your results.

PROCEDURE 15.2: CELLULAR RESPIRATION

1. Set two water baths to equilibrate, one at room temperature, and one at 37 °C.

2. Make five respirometers. Place a cotton ball in the bottom of each of five test tubes, and saturate each cotton ball with KOH solution, which will absorb CO_2 out of the system. Place nonabsorbent material over the saturated cotton.

3. On top of the cotton, place 15 germinated peas (or an amount that about half-fills the test tube) in two test tubes, an equal number of dried peas in two other test tubes, and an equal number of beads in a fifth tube as a control.

4. Carefully slide a glass pipet or dropper through the stopper, from the narrow side to the larger side for each test tube. Place the stoppers in the test tubes, with the pointed end of the pipet extending out of the test tube.

5. Without allowing the ends of the pipets to go under the water, allow the test tubes to equilibrate: place a germinating pea respirometer, a dried pea respirometer, and a bead respirometer in each of the water baths for 5 minutes/

6. Lay the respirometers flat under the water all at the same time. Wait 20 minutes, and look to see how much water has been pulled into the pipet. The more water that has been drawn in means the more oxygen has been used by the contents of the test tubes.

POST-LAB QUESTIONS

1. Under what conditions and with which item—germinated peas, dried peas, or beads—was the most water drawn into the pipet? Why is this?

2. If the pipets were graduated so that a measurement of water drawn in could be determined, what would the measurement of the bead tube be used for?

3. With a timer, would it be possible to determine the rate of oxygen consumption of the contents of the tubes? Why or why not?

PROCEDURE 15.3: ANTIOXIDANTS AND THE BROWNING OF FRUIT

1. Take five equally sized apple slices.

2. Label five paper towels: control, citric acid, ascorbic acid, acetic acid 0.1%, and acetic acid 1%.

3. Using tongs, dip one slice into each of the solutions, and lay it on the labeled paper towel. Dip the "control" slice in plain water.

4. Observe the slices for 20 minutes or more as they interact with the oxygen in the air.

POST-LAB QUESTIONS

1. Which of the three compounds was most effective in preventing the oxidation of the fruit slices?

2. The oxidative browning is facilitated by an enzyme, *polyphenol oxidase*. Given this piece of information, along with the information gained from the experiment, suggest two ways that lemon juice might prevent the browning of fruit.

3. Could the compounds identified in Question 2 above be used to prevent the rusting or corrosion of metals? Why or why not?